T0291644

CAMBRIDGE LIBRARY COLLECTION

Books of enduring scholarly value

Mathematical Sciences

From its pre-historic roots in simple counting to the algorithms powering modern desktop computers, from the genius of Archimedes to the genius of Einstein, advances in mathematical understanding and numerical techniques have been directly responsible for creating the modern world as we know it. This series will provide a library of the most influential publications and writers on mathematics in its broadest sense. As such, it will show not only the deep roots from which modern science and technology have grown, but also the astonishing breadth of application of mathematical techniques in the humanities and social sciences, and in everyday life.

The Growth of Physical Science

Sir James Jeans (1877–1946) is regarded as one of the founders of British cosmology, and was the first to suggest (in 1928) the steady state theory, which assumes a continuous creation of matter in the universe. He made many major contributions over a wide area of mathematical physics, but was also well known as an accessible writer for the non-specialist. This book, first published posthumously in 1947, is a detailed but very accessible survey of what began as natural philosophy and culminated in the mid-twentieth century as quantum physical science. Covering the earliest physical investigations of nature made by the various civilisations of Babylonia, Phoenicia and Egypt (a period covering 5000–600 BCE), through the remarkable mathematical and philosophical achievements of the ancient Greeks, to the ages of Newton and then Einstein, Rutherford and Bohr, this comprehensive history of the tremendous advancement in our understanding of the universe will still appeal to a broad range of readers.

Cambridge University Press has long been a pioneer in the reissuing of out-of-print titles from its own backlist, producing digital reprints of books that are still sought after by scholars and students but could not be reprinted economically using traditional technology. The Cambridge Library Collection extends this activity to a wider range of books which are still of importance to researchers and professionals, either for the source material they contain, or as landmarks in the history of their academic discipline.

Drawing from the world-renowned collections in the Cambridge University Library, and guided by the advice of experts in each subject area, Cambridge University Press is using state-of-the-art scanning machines in its own Printing House to capture the content of each book selected for inclusion. The files are processed to give a consistently clear, crisp image, and the books finished to the high quality standard for which the Press is recognised around the world. The latest print-on-demand technology ensures that the books will remain available indefinitely, and that orders for single or multiple copies can quickly be supplied.

The Cambridge Library Collection will bring back to life books of enduring scholarly value (including out-of-copyright works originally issued by other publishers) across a wide range of disciplines in the humanities and social sciences and in science and technology.

The Growth of Physical Science

James Jeans

CAMBRIDGE UNIVERSITY PRESS

Cambridge, New York, Melbourne, Madrid, Cape Town, Singapore,
São Paolo, Delhi, Dubai, Tokyo

Published in the United States of America by Cambridge University Press, New York

www.cambridge.org
Information on this title: www.cambridge.org/9781108005654

This edition first published 1947
This digitally printed version 2009

ISBN 978-1-108-00565-4

THE GROWTH OF PHYSICAL SCIENCE

The Alchemist

David Teniers the younger 1610–90

THE GROWTH OF PHYSICAL SCIENCE

BY

SIR JAMES JEANS

CAMBRIDGE
AT THE UNIVERSITY PRESS
1951

CAMBRIDGE UNIVERSITY PRESS
Cambridge, New York, Melbourne, Madrid, Cape Town, Singapore, São Paulo, Delhi

Cambridge University Press
The Edinburgh Building, Cambridge CB2 8RU, UK

Published in the United States of America by Cambridge University Press, New York

www.cambridge.org
Information on this title: www.cambridge.org/9780521744850

First published 1947
Second edition 1951
This digitally printed version 2008

A catalogue record for this publication is available from the British Library

ISBN 978-0-521-74485-0 paperback

PUBLISHERS' NOTE TO THE SECOND EDITION

This book was published posthumously in 1947. Sir James Jeans had seen and read the first proofs, but he died before the final touches could be added. Since publication a number of expert reviewers and readers have voluntarily supplied the publishers with notes of misprints in dates and names, and of places where the latest findings on matters dealt with had been overlooked. In the ordinary course Sir James Jeans would himself have revised his book in the light of these indications. As this was no longer possible the publishers have taken expert advice, and after collating the comments of several readers have put the preparation of the second edition into the hands of Mr P. J. Grant of the Cavendish Laboratory. They are grateful to all those who have helped in this way.

PREFACE

There are a vast number of detailed and comprehensive histories, both of general science and of special departments of science. Most of these are admirable for the scientific reader, but the layman sometimes cannot see the wood for the trees. I have felt no ambition (nor competence) to add to their number, but have thought I might usefully try to describe the main lines of advance of physical science, including astronomy and mathematics but excluding all points and side-issues, in language non-technical enough to be understood by readers who have no scientific attainments or knowledge.

I hope that such a book may prove of interest to the general educated reader, perhaps also to those who are beginning the study of physics, and possibly to students of other subjects who wish to know something of how physical science has grown, what it has done, and what it can do.

J. H. J.

CONTENTS

PLATES

CHAPTER I

THE REMOTE BEGINNINGS

(5000–600 B.C.)

WE look on helpless while our material civilisation carries us at breakneck speed to an end which no man can foresee or even conjecture. And the speed for ever increases. The last hundred years have seen more change than a thousand years of the Roman Empire, more than a hundred thousand years of the stone age. This change has resulted in large part from the applications of physical science which, through the use of steam, electricity and petrol, and by way of the various industrial arts, now affects almost every moment of our existences. Its use in medicine and surgery may save our lives; its use in warfare may involve us in utter ruination. In its more abstract aspects, it has exerted a powerful influence on our philosophies, our religions, and our general outlook on life.

The present book aspires to tell the story of how physical science has grown, and to trace out the steps by which it has attained to its present power and importance. To do this fully we ought to go back to the dim ages when there was no physical science, to the times before our cave-dwelling ancestry had begun to wonder why the night followed the day, why fire consumed and why water ran downhill.

This we cannot do. The early history of our race is hidden in the mists of the past, and the facts we should most like to know about its early days elude our search. We do not know, and probably never shall know, the people or peoples who first found that fire could be generated by friction, or first discovered the principles of the wheel, the sail and the lever. But we still have with us the implements and weapons that primitive man left behind him on the floors of his huts and caves, or buried with his dead; the pyramids of Egypt and the contents of their tombs; the buildings, the drawings and the domestic utensils of Susa, Erech, Ur and Knossos. From such fragmentary survivals the archaeologist can reconstruct something of the lives of these early peoples, and he finds that science of a primitive kind played its part in them.

The earliest evidence of any systematic interest in science comes from the civilisations which existed in the river-basins of the Euphrates and the Nile in the fourth and fifth millennia before the Christian era. Man was still in the 'neolithic' or new-stone age, but was about to enter the 'bronze' age* by learning how to harden his all-too-soft copper with an admixture of tin, and how to work the resulting alloy into tools and weapons. His artistic development at this time was well in advance of his scientific development, for he was already producing sculpture, pottery and jewellery, all of which showed skill of a high order.

The two civilisations just mentioned were geographically distinct, but can hardly have been entirely unrelated, since their cultures, their arts, and even their religions show certain features in common. A large mass of evidence† suggests that, sometime before 5000 B.C., a peaceful, artistic and highly talented race left their homes somewhere in central Asia and descended into Mesopotamia—the land 'between the rivers' Euphrates and Tigris, which is sometimes called the cradle of the human race, but might more accurately be described as the cradle of human civilisation. Mixing with the native races, they produced a new people, the Sumerians, who carried civilisation to a higher level than any of the constituent races had ever attained. They had considerable engineering skill, as is shown by the irrigation system they established in Lower Mesopotamia, probably in the fifth millennium B.C., as well as by their great temples and palaces. Even in the fifth millennium B.C., their craftsmen were already using the

* Bronze seems first to have been used in Crete, in about 3800 B.C., and in Egypt in the fifth dynasty (about 2800 B.C., or possibly earlier; all dates in this dim distant period are highly conjectural).

† *Cambridge Ancient History*, vol. I, chap. X–XII.

potter's wheel to make fine pottery, which they ornamented with a lustrous black paint, made by mixing brown haematite with an alkali salt and potash. Graves at Ur which date from about 3500 B.C. have yielded real art treasures of the finest workmanship, in gold and silver, copper and shell.

Some of the invaders stayed in Mesopotamia, but others seem to have passed on to Egypt, carrying a certain Sumerian influence with them. Here too a high level of civilisation was soon reached, as is shown by the very scientific determination of the length of the astronomical year. The Egyptians had taken their civil year to be exactly 365 days—12 months of 30 days each, together with 5 extra sacred or 'heavenly' days. But as the astronomical year, the precise period of the earth's revolution round the sun, contains rather more than 365 days, the two years did not keep exactly in step, and yearly natural events, such as the flooding of the Nile, marched steadily through the civil calendar. These floodings did not recur with sufficient regularity to fix the exact length of the astronomical year, and the Egyptians had to look for a more precise clock.

They found it in the risings of the stars in the east. Every star rises a few minutes earlier each day than it did the day before, so that every morning new stars can be seen which had previously been lost in the glare of the already risen sun. The day on which the star Sothis (our Sirius) first became visible was found to coincide very approximately with the beginnings of the Nile floods, and formed a sort of landmark which recurred every astronomical year. Here was a precise astronomical clock which ticked exact astronomical years. Observation showed that the first visible rising of Sothis advanced through the civil year at the rate of one day every four years, so that the astronomical year was seen to consist of 365¼ days, and the first visible rising of a star would return to its original place in the civil calendar after 1461 years. This period the Egyptians called the 'Sothic Cycle'. A new cycle is known to have commenced in A.D. 139, whence it is easy to calculate when the earlier cycles commenced. It seems likely that the Egyptians started their calendar with the cycle which commenced in about 4240 B.C., so that even in this early age, they had obtained an accurate knowledge of the length of the year through really scientific observation.

Before the beginning of the first dynasty (probably about 3400 B.C., but possibly much earlier), Egyptian artificers had been producing skilled work in copper, gold, alabaster and ivory. They had discovered that they could produce a decorative glaze by heating sand with potash or soda and a metallic oxide, and knew that they could colour this blue by adding a salt of copper to the melt.

They were already using writing materials—pens, ink and papyrus—and were employing an alphabet and a definite numerical system (p. 10, below). With these they kept a record of current events, including measures of the height reached by the Nile at its successive floodings. But perhaps the most striking evidence of their culture is the Great Pyramid of Gizeh, which was probably built about 2900 B.C. Its base is a perfect square, the sides of which run so exactly north-south and east-west that even the marking it out on the sands of the desert was no small achievement. Still more remarkable is the structure which stands on this base. Its faces are all perfectly plane—or were before their outer casings had been removed—and all have exactly the same slope of 51° 50′. Its 'bricks' are 2½-ton slabs of stone, fitted together so exactly that it is often impossible to insert the blade of a knife between them. The King's chamber, at the centre of the structure, is roofed by 56 slabs of 54 tons each, the placing of which must have called for geometrical and engineering skill of a very high order.

Farther to the east lay India and China, which certainly had highly developed civilisations 3000 years before Christ, and possibly highly developed sciences as well. The Chinese kept records of the appearance of comets from 2296 B.C. on, and the *Shu Chang*, a collection of documents of the same period, tells of the Emperor Yao ordering records to be kept of the dates of the longest and shortest days and of the equinoxes when the days and nights were equal in length. They may even have known how to predict eclipses at this time, for we read of two astronomers being put to death for having failed to do so. 'The blind musician has drummed, the mandarins have mounted their horses, the people have flocked together. At this time Hi and Ho, like wooden figures, have seen nothing, heard nothing and, by their neglect to calculate and observe the movements of the stars, have incurred the penalty of death.'

This suggests that astronomy must have been in a fairly advanced state in China, and it may have been equally so in India; we do not know. Fortunately, the question is not very important for our present inquiry, which is less concerned with the sowing of the seed than with the fruiting of the tree. Our main study will not be the origin of physical science, but its growth, and this hardly got under way until the sixth century B.C. Then it started in Ionian Greece, the ragged fringe of coastline and islands which forms the westerly edge of Asia Minor, and gradually spread from here, first to the Greek mainland and thence to the rest of Europe.

Greece was still a new civilisation. To its east lay the mature civilisations of China, India, Persia, Mesopotamia, Phoenicia, Crete and Egypt; to its west lay lands still untouched by civilisation—the wild, barbaric lands of the setting sun. Science, like the rest of civilisation, dawned on these lands from the east. Ideas and knowledge began to flow from the old civilisations of the east to the new civilisations which were springing up in the west, the flow being fostered by trade and occasionally expedited by colonisation or military conquest. India and China contributed to western science only through the intermediary of the near east, so that we shall not go far wrong if we disregard these remoter eastern civilisations and confine our attention to the nearer which formed direct stepping-stones into Europe. Foremost among these were Mesopotamia (or Babylonia, as we ought to designate it by now), Egypt and Phoenicia; let us look at the contributions which these civilisations were able to make by the sixth century B.C.

BABYLONIA

One of the greatest scientific achievements of the Babylonians was their numerical system and method of reckoning. Like most primitive peoples, they first used a simple decimal system—i.e. they counted by tens—probably, as Aristotle suggests,* because the human body has ten digits.

But, as many peoples have found, a decimal system is not ideally convenient. It has often been remarked that arithmetic might have been much simpler if men had possessed twelve fingers instead of ten. Then we should probably have counted by twelves and used a duodecimal system. As 12 can be exactly divided by 2, 3, 4 and 6, this has the advantage that awkward fractions are less likely to appear—such as $33\frac{1}{3}$ % and 6·25 which appear in the decimal system because 10 cannot be divided by 3 or 4. Yet even a duodecimal system is not perfect, since 12 cannot be divided by 5. The later Babylonians tried to combine the advantages of both systems by using a sexagesimal system, in which the larger unit consisted of 60 smaller units; and 60 can be divided exactly by no fewer than ten factors—2, 3, 4, 5, 6, 10, 12, 15, 20 and 30. They employed this system in tables which date back to about 2000 B.C. It has proved so convenient in practice that it still survives in the 60 minutes of the hour, and the 60 seconds of the minute, as well as in the corresponding subdivisions of angles.

The Babylonians combined this sexagesimal system with a scheme of notation which was 'positional' in the sense that the value of a symbol depended on the place it occupied in a number, an advantage which was conspicuously and dis-

* In his *Problems* he asks: 'Why do all men, whether barbarians or Greeks, count up to ten, and not to any other number? It cannot have been chance, for what is always and universally done is not due to chance....Is it because men were born with ten fingers, and so use this number for counting everything else as well?'

astrously lacking in the much later Greek and Roman schemes of enumeration. In our modern notation, 123 stands for $1\times(10)^2+2\times(10)+3$, the numerals denoting hundreds, tens or units according to their position. In the same way, for the Babylonians ′ ″ ‴ stood for $1\times(60)^2+2\times(60)+3$.* They had a similar notation for fractions. Just as we write 1·23 to indicate $1+\frac{2}{10}+\frac{3}{(10)^2}$, so they wrote ′ ξ ″ ‴ to indicate $1+\frac{2}{60}+\frac{3}{(60)^2}$. They did not pass this system of notation directly on to Europe, but it was probably the origin of the decimal Hindu-Arabic system which ultimately came to the western world through the Arabs (p. 105), and is used by the whole world to-day; it is not known when or how the change was made from sixties to tens.

They sometimes went even further in the same direction, dividing their league into 180 cords and the cord into 120 cubits. They also divided the complete circle into 360 degrees. Some think they did this by taking the angle of the equilateral triangle (60°) as their fundamental unit, and dividing this into the usual 60 subunits. Others think that astronomical considerations may have come into play. When the early Babylonians first tried to measure the number of days in the year, they would find it was about 360. More than 2000 years before Christ they agreed to call it 360 as an approximation, dividing their year into 12 months of 30 days each, and inserting extra months now and then as needed to prevent the calendar running away from the seasons. At a later date, they traced the Zodiac—the path in which the sun, moon and planets appear to travel across the sky—and divided it into 12 equal divisions, so that the sun moved through one every

* In place of our ten so-called Arabic numerals (see p. 110) 0, 1, 2, ..., 9, the Babylonians used only two symbols—the wedge-shaped ′ to denote unity and ⟨ to denote 10; we can almost imagine these to have represented a finger and two outstretched hands. For instance, they wrote ⟨‴′ for our 14.

month. It was now natural to divide each of these into 30 parts, one of which the sun would traverse every day, and the complete circle would now be divided into 360 equal units.

There is evidence that the Babylonians not only named the twelve divisions of the Zodiac, but also divided the northern sky into 'constellations', or groups of stars, and gave these their present names. They did not travel into the southern hemisphere, and so could never see the stars surrounding the south pole of the sky; here the constellations have modern names, such as the 'clock' and the 'telescope'. But the constellations of the northern sky carry the names of the legendary figures and heroes of antiquity, suggesting that they were grouped and named in ancient times.

The earth wobbles as it rotates (p. 92), so that the portion of the sky which can be seen from any part of the earth's surface is continually changing; that part in which the constellations bear ancient names is the part which could be seen from about latitude 40° N., in about the year 2750 B.C., and this is thought to suggest that these constellations were grouped and named by the Babylonians of some such date. They are practically identical with our present-day constellations of the northern sky. The Chinese group and name their constellations differently, showing that our constellations did not come to us from the Chinese.

The early astronomers did not know how to measure small fractions of a day with any precision; no one did, until Galileo discovered the principle of the pendulum clock early in the seventeenth century (p. 149). Nevertheless, some two or three millennia before the Christian era Babylonian priests were recording planetary motions with fair accuracy, especially those of Venus. One temple is said to have possessed a library of tablets of such observations which dated from before 3000 B.C., while a later set commencing about 747 B.C. proved very valuable to later generations of astronomers. By the seventh century B.C., the movements of the heavenly bodies were being regularly recorded at a complete system of observatories and reports were being sent to the king, who seems to have controlled both the observatories and the calendar.

The Babylonian astronomers of more recent times knew enough of astronomy to be able to predict eclipses. The sun is eclipsed whenever the moon comes directly between it and the earth, so that if the sun, earth and moon all moved in one plane, there would be an eclipse every lunar month. Actually the three bodies move in different planes, so that eclipses in a given region recur after a period of 233 lunar months, which is equal to 18 years and 11⅓ days. This period of time is known as the 'Saronic cycle' or, more briefly, the 'Saros'. Through knowing of the Saros, the Babylonians were able to predict eclipses as long ago as the sixth century B.C.

Later still they made some amazingly accurate measures of other astronomical periods. In particular, the following estimates of the length of the lunar month have survived:*

Naburiannu (about 500 B.C.)	29·530614 days
Kidinnu (about 383 B.C.)	29·530594 days
True value	29·520596 days

Precise knowledge of this kind carried with it a limited power of foreseeing and predicting the astronomical future, and this no doubt accounts for the phenomenal vogue of astrology in Babylonia, and the astonishing prestige which the Babylonian astrologers enjoyed throughout the ancient world. For if a student of the sky could foretell the movements of the sun, moon and planets, and if—a belief which the astrological fraternity were careful to inculcate—the movements of these bodies influenced human affairs, then the astrologer could obviously save his clients from harmful influences and show them how to turn beneficial situations to their greatest advantage.

Geometry, too, seems to have had a period of brilliance in Babylonia. Recently deciphered tablets of about 1700 B.C. show that the Babylonians of that time were acquainted with

* Sir T. L. Heath, *Greek Astronomy*, p. liii.

the famous 'Theorem of Pythagoras' (p. 28), which the Greeks rediscovered in the fifth century B.C., and even knew how to find sets of integral numbers (e.g. 3, 4 and 5) such that triangles having sides of these lengths would be right-angled. The Greeks were great geometers but, in this one particular instance at least, the Babylonians were a good thousand years or so ahead of them.*

Other tablets of the same period show that the Babylonians of this time were skilled in arithmetical calculations. They contain a number of tables for the solution of problems which lead to quadratic equations, such as the determination of two numbers, the sum and product of which are known. There are also tables giving the power to which an assigned number must be raised so as to yield another assigned number, these having apparently been used for the calculation of compound interest; indeed, two examples of such calculation are appended, the rates of interest incidentally being 20 and $33\frac{1}{3}$ per cent.!

Egypt

Egypt and Babylonia had been in such close commercial and cultural contact from the earliest times that they inevitably had much in common. The Egyptians, like the Babylonians, had a good decimal notation for whole numbers,† but they failed with fractions. It was their practice, in which the Greeks followed them until at least the sixth century A.D., to express all fractions (with the single exception of $\frac{2}{3}$) as a sum of aliquot parts—i.e. of fractions each having unity as its numerator. For instance, they thought of $\frac{3}{4}$ only as $\frac{1}{2}+\frac{1}{4}$.

Our knowledge of their arithmetical methods comes largely from a papyrus which forms part of the Rhind collection in the British Museum. This dates from about 1650 B.C., but is

* *Mathematical Cuneiform Texts*, Neugebauer and Sachs (New Haven, 1945). See also Sir T. L. Heath, *Manual of Greek Mathematics*, p. 96.

† Units and tens were represented by | and ∩, in place of the Babylonian ′ and ⟨, while there were other symbols for hundreds, thousands, and so on, up to millions.

only a copy, by a priest Ahmes, of an earlier papyrus which would seem from internal evidence to have been written many centuries earlier. It records the resolution of a great number of fractions into a sum of aliquot parts, the original numerator always being 2; as, for instance,

$$\frac{2}{97} = \frac{1}{56} + \frac{1}{679} + \frac{1}{776}.$$

But no rules are given for effecting such resolutions, and the whole treatise seems to be a mere compendium of results obtained by repeated trials. We get the impression of a plodding, unimaginative race.

The Egyptians multiplied by a method which is said to have been in use in Russia until quite recently. The 'multiplicand'—the number to be multiplied—is first doubled, then redoubled, and so on, thus providing a table giving 2, 4, 8, 16, ... times the multiplicand. From this table they took the entries needed to give the required result and added them together. For instance, to multiply by 13, the Egyptian arithmetician would add together the entries for 1, 4 and 8 times the multiplicand.

They had a simple procedure for finding which entries are needed. Suppose we wish to multiply 117 by 13. We first write down 13 and 117 on the same line, as in the margin. We next divide 13 by 2, disregard the unit remainder, both here and wherever else it occurs, and write down the quotient 6 under 13. At the same time, we multiply 117 by 2, and write down the product 234 under 117, thus completing the second line. We repeat the process to obtain a third line and continue until the first entry is reduced to 1. We now strike out all lines in which the first entry is even—in this case the second line alone—and add all that remains in the second column, as has been done in the margin.* The sum 1521 is the product we need. This method reduced all

13	117
~~6~~	~~234~~
3	468
1	936
	1521

* The mathematician will see the reason for this if he notices that odd and even numbers in the first column correspond to digits 1 and 0 when the multiplicand is expressed in the scale of 2.

multiplication of integers to a series of multiplications by 2; fractions could be multiplied by 2 by using the table of resolutions mentioned above.

In general astronomy the Egyptians were far behind the Babylonians; they did little more than record the appearance of the sky on various occasions, and even this was for worship rather than for study. There seems to have been no curiosity as to why things happened in the sky as they did—only a perfectly unimaginative record of the happenings.

On the other hand, Egyptian geometry was probably well in advance of Babylonian. This is hardly surprising. The yearly flooding of the land by the waters of the Nile involved an annual return to the Sisyphus-like task of mapping out fields, and this gave a special importance to the study and practice of geometry. The Rhind papyrus contains a number of rules for measurement, as well as some geometrical information of a more abstract kind, but difficulties of language often obscure the meaning. We cannot, for instance, tell whether the area of a triangle is said to be half the base multiplied by the height, or half the base multiplied by the side. The former is of course correct, the latter incorrect, but the two become almost the same thing when the triangle is very high and narrow, as it is in the diagram shown in the papyrus.

A more recently discovered papyrus, the Moscow papyrus* of the Twelfth Dynasty (probable date about 1800 B.C.), shows a far more extensive knowledge of abstract geometry. For instance, it contains a correct formula for the volume of a truncated pyramid, i.e. the piece of a pyramid cut off by a plane parallel to its base, like a partly finished stone pyramid. It is astonishing that the correct formula should have been discovered so long ago. The interesting question is, how was it discovered? There are sharp differences of opinion about this, one view crediting the Egyptians with the necessary knowledge of algebra while another points out that the formula may be derived from a simple averaging process. This seems on the whole the more likely explanation.

* First published in 1930.

But the true greatness of Egypt was not in its mathematics; it lay rather in medicine. Carvings dating from about 2500 B.C. depict a surgical operation in progress, while the Ebers papyrus of about 1600 B.C. contains a complete treatise on the preparation of drugs and therapeutic essences, and another, the Edwin Smith papyrus, is a really scientific treatise on surgery. Medicine and surgery were the only sciences outside geometry and engineering in which the Egyptians seem to have excelled, and neither in Egypt nor in Babylonia was there anything worth calling physical science.

Phoenicia

Strabo tells us that the Phoenicians paid special attention to the sciences of numbers, navigation and astronomy. We may safely believe the statement. They could hardly have become the great trading power of antiquity unless they had possessed considerable numerical aptitude, nor the greatest seafaring nation of their time unless they had studied navigation and astronomy. Little evidence of this has survived; any documents there may have been disappeared without even, so far as we know, being quoted. But we read that in the sixth century B.C. Thales (p. 58) advised the Greeks to adopt the Phoenician practice of finding the north from the Little Bear instead of from the Great Bear, as was usual with them. They do not seem to have followed his advice, for some six centuries later we find a minor Greek poet, Aratus writing: 'It is by Helice [i.e. the Great Bear] that the Achaeans on the sea judge where to direct the course of their ships, while the Phoenicians put their trust in the other [i.e. the Little Bear] as they cross the sea. Now Helice is bright and easy to note, appearing large from earliest nightfall; the other is smaller yet better for sailors, for the whole of it turns in a lesser circuit [i.e. it is nearer to the north pole], and by it the men of Sidon steer the straightest course.'*

It is significant that the two greatest scientists of early

* Sir T. L. Heath, *Greek Astronomy*, p. 113.

Greece, Thales and Pythagoras, were both reputed to have been of Phoenician extraction, as was also Euclid the geometer and Zeno the philosopher, although many doubt these claims.

Greece

The study of physical science is ultimately a search for law and order in phenomena, so that it cannot flourish without the tools needed for the discovery and discussion of whatever law and order there may be. The fundamental tools needed in the physical sciences are arithmetic, geometry and techniques for the measurements of time and space.

Now these tools seem to have been available in early Egypt and Babylonia, and possibly in Phoenicia also, in ample measure, relative to the needs of the time. But no serious scientific use was made of them until many centuries later, and when a true scientific spirit first began to flourish, it was neither in Egypt nor in Babylonia, but in a small Greek colony on the shores of the Aegean Sea. There was no complete break with the past, but tender plants seemed to acquire a new capacity for growth—as though the fresh soil of Greek civilisation provided some new factor that had been lacking in the older civilisations. What, then, was this new factor? Partly perhaps the liberation of knowledge from the priesthood and its transfer to the laity. For, as Farrington writes:* 'The organised knowledge of Egypt and Babylon had been a tradition handed down from generation to generation by priestly colleges. But the scientific movement which began in the sixth century among the Greeks was entirely a lay movement. It was the creation and the property, not of priests who claimed to represent the gods, but of men whose only claim to be listened to lay in their appeal to the common reason of mankind.' More generally, it was perhaps that special kind of intellectual curiosity which impels men to try to understand rather than merely to know.

* *Science in Antiquity*, p. 36.

The Egyptians, as Plato said, had no such love of knowledge as the Greeks had; their passion was rather for riches and material prosperity. They had accumulated masses of particular and isolated facts, but had no idea of letting one fact point the way to another. Knowledge was a matter of revelation, a gift from the gods, and it was not for man to try to discover what Thoth* (Hermes) had left untold. And so we read of the priestly watchers of the stars standing on their pylons night after night to record the positions of the planets, but we hear of no attempt to discover the laws governing their motions.

The Babylonians were influenced by their astrological success, which urged them to perfect the very lucrative arts of foretelling the astronomical future, but again we hear little of their trying to increase their knowledge from sheer intellectual curiosity, or of using what knowledge they had for any purpose except astrological gain. Knowledge had been piling up in Egypt and Babylon, and perhaps in Phoenicia also, but the quest of knowledge for its own sake made but little appeal until the Greeks came.

Who, then, were these Greeks who showed these new capacities and interests, and so could weld the raw material of disconnected facts into a science? Where did they come from, and whence did they draw their intellectual powers?

We do not know; it is one of the great unsolved mysteries of history. The great civilisations of antiquity—the Indian, the Chinese, the Persian, the Egyptian, the Minoan civilisation in Crete and the Babylonian in Mesopotamia—all had been established for thousands of years before the Greeks appeared, and each had its own distinctive and well-marked characteristics. The newer Greek civilisation did not bear the stamp of any of them. It was something fresher and younger, and it was certainly different. The first clear picture we have

* In Plato's *Phaedrus*, Socrates says that he had heard that the Egyptian god Thoth was the first to invent arithmetic, the science of calculation, geometry and astronomy.

of it is in the Homeric poems, which are thought to have assumed their present form in the ninth century B.C., but probably describe Greek civilisation of about two centuries earlier. They tell us of an eager and joyous race, living with their bodies rather than with their minds, and untroubled by any doubts about the world in which they found themselves; their ideal was to hold both hands before the fire of life, and enjoy it to the full while it lasted. Apart from mentioning a few stars by name, the poems show no acquaintance with physical science of any kind, and there is nothing in them to suggest those powers of abstract thought and intellectual curiosity which were to come to such splendid fruition a few centuries later.

Yet on the artistic side, many have noticed a resemblance between the new Greek civilisation and the older Minoan civilisation which was centred in Knossos of Crete *c.* 3500–1500 B.C.; they find the same conception of beauty and the same sense of form in both, the same exquisite craftsmanship and the same care for detail. Scholars are still unable to read the Minoan script, but some think that the later Greek civilisation must have owed a great deal to the earlier Minoan. The position of Knossos made it a natural trade centre, and it may well have received ideas as well as goods from the east, and passed them on to the west.

Even so, this does not tell us where the Greeks themselves originated. Many scholars have imagined invading warriors —Homer's Achaeans—entering Greece somewhere about 1400 B.C., possibly armed with weapons of iron which speedily vanquished the primitive weapons of the natives, the Pelasgians. Some think they came from western Asia or the Russian steppes; others think they came from the Danube basin or northern Europe. Some think that the main torrent of invaders impinged on the Greek mainland, while subsidiary streams passed on to settle on the islands and coasts of the Aegean Sea—the westerly fringe of Asia Minor—where they formed the settlements of Ionia to the north, and Doria to the south. With them they brought their tribal gods, the sky-and-rain god Zeus, mountain-dwelling and thunderbolt-hurling, together with his retinue of sons and daughters—Apollo, Athena and the rest. These were soon accepted by the Greeks as their official gods, but had to share their sovereignty with other gods who were already established in Greece, and came in direct descent from the fertility-gods of yet more primitive tribes.*

In any case, it seems a safe conjecture that the Greeks were a mixed race, and their civilisation a blend of ingredients from many sources. History provides many instances of a successful new civilisation emerging from an admixture of invading conquerors with a more primitive native race; as when tin is mixed with copper, something new results which is better than either ingredient. And so it may have been with Greece.

Somewhat suddenly, we encounter the distinctly Greek intellect, and with it the first group of scientists that we can recognise as such. The time was the sixth century B.C., when, as Herodotus says: 'The Greek race was marked off from the barbarians, as more intelligent and more emancipated from silly nonsense.' The place was Ionia, and more especially Miletus, the largest city in Ionia and perhaps in all Greece, although its population can hardly have been more than 10,000. It was a great centre for trade, especially with Egypt, and as it had founded more than sixty daughter-cities on the shores of the Mediterranean, it must have enjoyed a constant interchange of ideas with other Mediterranean countries. Pottery which has been excavated there shows that it existed in Minoan times; by the middle of the sixth century B.C. it had become pre-eminently the centre of Greek culture, a sort of focal point through which all rays of intellectual light were likely to pass on their way from east to west.

* See Gilbert Murray, *Five Stages of Greek Religion.*

CHAPTER II

IONIA AND EARLY GREECE

(600–320 B.C.)

IN the present chapter we examine the first three centuries of Greek scientific progress; our period begins with the earliest impact of oriental scientific ideas on Ionian Greece, and ends with the conquest of Greece by Alexander the Great (332 B.C.), the death of Aristotle (322 B.C.), a general decline of science and art in Greece, and the foundation of the City of Alexandria and of its university (323 B.C.), which was to be the intellectual centre of the world for many generations to come. In brief, we study Greek science in the period of Greece's intellectual greatness.

This science was almost entirely mathematical. The Greeks had nothing of our elaborate equipment of laboratories and observatories. Indeed, their equipment was limited to their own brains, but these were of the very best; just as Aeschylus and Sophocles exhibit mental powers comparable with those of Shakespeare, so Archimedes and Aristarchus exhibit powers comparable with those of Newton. Thus they could attack their various problems only by reflection and contemplation, aided at most by a minimum of observation, and when physics and astronomy creep in, it is in the form of philosophical speculation rather than of true science as we understand it to-day.

It will be convenient to discuss the early Greek mathematics, physics and astronomy separately, and in this order.

GREEK MATHEMATICS

THE IONIAN SCHOOL

THALES. First and foremost among the Greek mathematicians stands Thales, who was born in Miletus in or about 624 B.C., and lived until about 546 B.C. Herodotus says he was

Phoenician extraction, but other accounts say he came of a noble Milesian family.

Intellectually he was a giant, and, as with so many of the great figures of science, his talents were as varied as they were numerous. 'Statesman, engineer, man of business, philosopher, mathematician and astronomer, he covered almost the whole field of human thought and activity.'* Like many thinkers, he got a reputation for living in a world of his own; Plato records a story of his walking into a well while he was looking at the stars, and being 'rallied by a clever and pretty maidservant from Thrace' for being so eager to know what went on in the sky that he could not notice what was happening at his own feet. Notwithstanding this lapse, he appears to have been particularly shrewd in practical affairs. Aristotle relates that one year when the olive crop promised to be particularly abundant, he made a 'corner' in olive presses, buying all he could, and then reaping a fortune by letting them out at his own price. He was clearly an engineer of some capacity, for he was commissioned to get the army of Croesus across the river Halys on dry feet. He did this by making an artificial river-bed by the side of the natural one; after the army had walked across the old and now dry river bed, the water was turned back into its old course. And we read more than once of his intervening effectively in politics.

As it was largely through his activities that the scientific spirit first entered into Greece, we should especially like to know where and how he acquired his interest in science, but this information is lacking. There may have been some Babylonian influence, for it is possible that during his travels (no doubt partly connected with his business activities) he studied as a pupil of one of the Babylonian priests. On the other hand, although we know that Thales travelled a great deal, particularly in Egypt and Babylonia, we are definitely told†

* Sir T. L. Heath, *Manual of Greek Mathematics*, p. 81.

† By Diogenes Laertius, quoting Hieronymus of Rhodes.

that he never had any teacher except when he associated with the priests in Egypt.

However this may be, a man of such wide and varied interests could be trusted to assimilate any scientific ideas he met with on his travels; there were few enough in those days. He is likely to have acquired some geometrical knowledge in Egypt, and to have learned in Babylon of the Saronic cycle and the Babylonian method of predicting eclipses. Herodotus records that when he returned home, he gained a great reputation in Miletus by predicting an eclipse of the sun.* It occurred during a battle between the Medes and the Lydians, and the darkness was so complete that the fighting had to stop. This was thought to show that the gods wished the war to end, and a peace was arranged. Thus not only the eclipse but the prophecy also was brought into prominence, and in 582 B.C. Thales was declared one of the 'seven wise men' of Greece—the only philosopher in a crowd of politicians; Plutarch, writing about A.D. 100, says he was the only one of the seven 'whose wisdom stepped out in speculation beyond the limits of practical utility'.

None of his writings have survived; we know them only at third hand. Just about a thousand years after his death, the Athenian philosopher Proclus (p. 37) wrote a *Commentary on Euclid*, which commenced with a brief summary of Greek mathematics up to the time of Euclid.† This tells us that Thales went to Egypt, and introduced the study of geometry from there into Greece, and that he was not interested in it solely for its practical applications, but also 'as an abstract

* Recent investigations cast some doubt on the whole story, but I have recorded it in the form in which it is told by Herodotus and repeated by innumerable other historians. If the story is true, the eclipse would probably be that of 28 May 585 B.C., although Eudemus, in his *History of Astronomy*, says it was about the fiftieth Olympiad, 580–577 B.C.

† The authorship of this summary is unknown. It was formerly known as the Eudemian Summary from a probably mistaken belief that it had been written by Eudemus, a pupil of Aristotle, and was an extract from his great *History of Geometry*. See Sir T. L. Heath, *A History of Greek Mathematics*, I, 118.

deductive science, based on general propositions'. It credits him with a knowledge of the following four propositions:

(1) Any diameter of a circle divides it into equal parts.

(2) The angles at the base of an isosceles triangle are similar (fig. 1).*

(3) When two straight lines cross one another, the opposite angles are similar (fig. 2).†

(4) When the base of a triangle is given, and also the angles at its ends, then the triangle is completely determined.

Plutarch credits Thales, at least by implication, with the further knowledge that when two triangles are of the same shape (i.e. when their angles are the same), their sides are

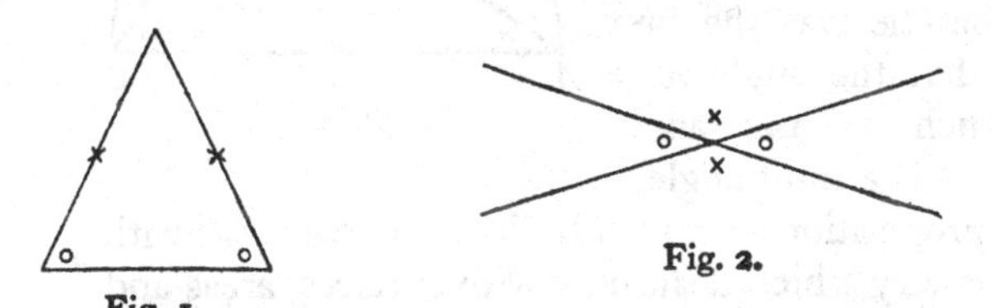

Fig. 1.

Fig. 2.

proportional. For he says that Thales measured the height of an Egyptian pyramid by comparing the length of its shadow with that cast by a stick of known length. If, for instance, a 6 ft. shadow was found to be cast by a 3 ft. stick, then a 600 ft. shadow would be cast by a pyramid 300 ft. in height. Plutarch adds that this method of measurement greatly impressed the Egyptian King Amasis, who was present. But earlier writers, Hieronymus and Pliny, say that Thales chose the precise moment when the shadow was equal to the height of the object which cast it. If this was all, Thales may not have been acquainted with the more general

* An isosceles triangle is one in which two sides are equal; in fig. 1, the sides marked × are equal, so that the angles marked o are similar. The use of the word similar, instead of equal, suggests that Thales did not think of an angle as a magnitude, but rather as a shape formed by lines.

† Thus in fig. 2, the angles marked × are similar, as also are those marked o.

proposition, nor even with the rather difficult idea of proportionality. On the other hand, Proclus says that Thales was able to determine the distances of ships out at sea, and that his method involved the theorem of proportionality; the details of the method are not known.

Thales is also credited with another proposition which he must have thought important, for he is said to have sacrificed an ox to the immortal gods to celebrate its discovery. Pamphila, who wrote in or about the reign of Nero (A.D. 54–68), records it in the form, 'Thales was the first to inscribe a right-angled triangle in a semicircle', meaning apparently that he was the first to discover that the angle in a semicircle—such as the angle *ADB* in fig. 3—is a right angle.

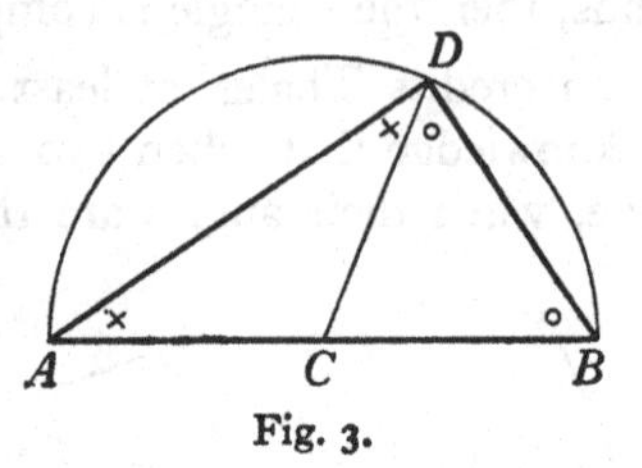

Fig. 3.

All these propositions deal with lines, in contrast with Egyptian geometry which dealt only with surfaces, areas and volumes; we may say that Thales was the creator of the geometry of lines. Further the propositions of Thales announced abstract universal truths, in contrast with the propositions of the Egyptians which were concerned with practical measurements; Thales established abstract geometry as a science.

We do not know how Thales reached his various results. So long as geometry proceeds by purely deductive methods, nothing can come out of it that has not been previously put into it in the form of assumptions. It would be interesting to know exactly what assumptions Thales made to arrive at his propositions. Some are of course so simple that the question hardly arises; for instance, we can *see* that a circle is bisected by its diameter as soon as we double it back on itself with a diameter as a hinge. But the proposition about the angle in a semicircle is less obvious. It is easy to prove it deductively if it is known that the sum of the three angles of a triangle is equal to two right angles, but otherwise not. And Thales is hardly likely to have known this; he did not think of an angle as a magnitude, so that the idea of adding angles was foreign to his thought; also Proclus definitely attributes the theorem to the Pythagoreans, who came some 50 years after Thales (p. 27). On the other hand, Thales may well have known, as a matter of fact, that the two diagonals of a rectangle are equal and bisect one another; this is the kind of relation that jumps to the eye on inspecting a tiled floor, besides being obvious from the consideration that there can be no reason for one semi-diagonal being longer than any other (fig. 4). If Thales had ever noticed this, he would see at once that a circle could be drawn through the four corners of any rectangle, and the truth of the theorem would become obvious.

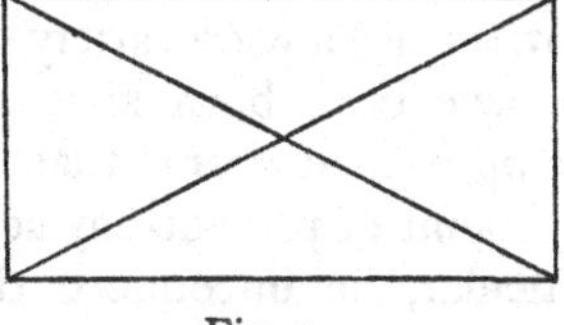

Fig. 4.

Many of Thales's proofs may have been of this semi-intuitive kind; indeed, Proclus tells us that he 'discovered many propositions,...his method of attack being in some cases more abstract, in others more observational' (αἰσθητικώτερον).

Most of his 'discoveries' were so rudimentary that any schoolboy of to-day would dismiss them as obvious. But this is only to say that Thales stood at the very fountain-head of European geometry, where he turned the stream of discovery into truly scientific channels, so that the tracing out of the history of geometry, and thence of mathematics and science in general, is merely tracing the course of this stream. The major practical achievements of physical science—the electric current, the telegraph and telephone, the aeroplane and the motor-car, radio and television—all are of western origin, and if we follow them back to their ultimate source, we find that they all trace back to the stream of knowledge started by Thales of Miletus.

ANAXIMANDER. Thales found an obvious successor in his fellow-citizen and friend Anaximander, who was born about 611 B.C., and lived until about 545 B.C. Suidas tells us that he wrote a book on geometry, so that he seems to have carried on the geometrical tradition of Thales, but he was apparently more interested in astronomy, geography and the general problems of philosophy. Apart from his geometry, he is said to have written only one book, *On Nature*, which appeared shortly before his death. Theophrastus* gives an account of him and his doctrines which conveys an impression of great mental powers and a wide variety of interests. Sometimes he seems to have been born 2000 years before his time; he certainly approached nearer than any of his contemporaries to the viewpoint of present-day science.

In particular, he introduced the idea of evolution into science. Hippolytus† tells us he distinguished 'the three stages of coming into being, existing and passing away'. He attributed all change to motion, and maintained that there are an infinite number of worlds, all in motion, 'since without motion there can be no coming into being or passing away'. He also introduced evolutionary ideas into biology, saying that living things first originated in slime which was evaporated by the sun; they had prickly coverings at first, but subsequently moved to drier places. Man, he thought, was born inside a fish, and was at first like a fish. He is different from all other animals, which find food for themselves soon after birth, for he needs long nursing; if man had originally been as he is now, he could not have survived.

Besides all this, he was the first geographer to attempt a complete map of the inhabited parts of the earth's surface. He also showed how time could be measured by a sort of primitive sundial in which a shadow was cast by a vertical stick, but the Babylonians had used this device before him.

After his death, the Milesian school gradually transferred its interest to philosophy, and finally came to an end somewhere

* *Physical Opinions.* † *Refutatio omnium Haeresium.*

about 400 B.C. We need not concern ourselves with it any more; if we wish to watch the development of science, we must leave Miletus with one of its most distinguished citizens who carried the torch of geometrical learning still farther to the west. We must give our attention to the mysterious and mystical figure of Pythagoras, and the school he founded.

The Pythagorean School

PYTHAGORAS. We know little of his life, birth or death. His birthplace was Samos in Ionia and, like Thales and Euclid, he is said to have been of Phoenician extraction, but the claim is suspect. The only certain date in his life is 530 B.C., when he left Samos to found a school in Croton, a Dorian colony in southern Italy. He was then young enough for his mother to be still alive (for he took her with him), and old enough to leave his birthplace on political grounds, whence it is generally assumed that he may have been born about 570 B.C. Iamblichus* says that Thales was so impressed by Pythagoras's ability that he imparted his own store of knowledge to him, and advised him to go and study with the Egyptian priests. This he did, studying astronomy and geometry from the age of 22 to 44, after which he lived in captivity in Babylon for 12 years and 'attained to the highest eminence in arithmetic, music and other branches of knowledge'. But it is difficult to fit all this together into a consistent biography.

In Croton he founded a sort of brotherhood of learned men, the members of which possessed all things in common—knowledge, philosophy and goods—ordering their lives by a common moral code, and forming a body rather like a modern religious order. Its members preached and practised strict self-control, temperance and purity, living simple ascetic lives and avoiding animal food because they believed that the beasts were akin to man—one of the few instances of consideration for the animal kingdom that we meet before the modern era; indeed Pythagoras is quoted with Empedocles

* *De Vita Pythagorica.*

as the founder of this branch of morality.* In brief, the Pythagoreans hoped, through abstinence, discipline and religious ceremonies, to purify the soul, free it from the wheel of birth and fit it for the life after death. For they regarded the body merely as a temporary prison for the soul, Pythagoras himself advocating the doctrines of immortality and transmigration of the soul, having learned both from his teacher Pherecydes of Syros. Pythagoras wrote: 'When we live, our souls are dead and buried in us, but when we die, our souls revive and live.'

In practical affairs the Pythagoreans aimed at a moral reformation of society, and this led to their undoing. Their advocacy of government by the best men, a true aristocracy in the most literal sense of the word, brought them into frequent conflict with the democratic mob who finally, in or about 501 B.C., killed many of them and burned their houses, while their founder fled to Tarentum. Accounts differ as to how the affair terminated, but the society seems to have ended a troubled existence somewhere about the middle of the fourth century B.C.

The daily occupation of the brotherhood was the acquisition of knowledge, and this they shared only with one another; anyone who divulged it was thought worthy of death. We read of two Pythagoreans being drowned at sea, and in each case it was said to serve him right; one, named Hippasus, had boasted that he had discovered a new regular solid, the dodecahedron (p. 34), while the other had disclosed the incommensurability of $\sqrt{2}$ (p. 32).

This habit of secrecy makes it very difficult to say how much the Pythagoreans achieved in science, and impossible to assign results to their individual authors. Our most useful guide is an account of the Pythagorean philosophy and teaching which the astronomer Philolaus (p. 127) wrote some ninety years after the death of Pythagoras. Nothing of the original book has

* Lecky, *History of European Morals*, II, 166; *Cambridge Ancient History*, IV, 576.

survived, but parts of it are described in the so-called Eudemian fragment (p. 20). Plato is said to have drawn upon this book for his only scientific dialogue, the *Timaeus*. Proclus tells us that Pythagoras 'transformed the study of geometry into a liberal education', while Aristoxenus says he 'advanced the study of arithmetic, and took it out of the region of commercial utility'.

The Pythagorean arithmetic was much concerned with the mystical properties of integral numbers. We all know how superstition can link up ideas with numbers; 3 and 7 may be sacred, 13 unlucky, 666 the number of the beast, and so on. The ideas of the Pythagoreans have been recorded by Aristotle. They associated the number 1 with a point, 2 with a line, 3 with a surface and 4 with a space. This was simple enough, but 2 was also associated with opinion because both are 'unlimited and indeterminate', and also with femininity, for reasons unspecified. Three was not only associated with the idea of surface, but also with masculinity. Then 4 was associated with justice because $4=2\times 2$, and so is the product of two evenly balanced factors. Next 5 is associated with marriage, because it results from the union of the male 3 and the female 2, and 7 with virginity because it has no factors. There were also ten fundamental oppositions associated with odd and even numbers, such as the finite and the infinite, the one and the many, the right and the left and so on. It seems unbelievably futile to-day, but the Pythagoreans thought it would provide the key to the universe. Aristotle says they thought that numbers not only expressed the form of the universe but also its very substance. Just as, in a later age, Plato thought that the world consisted primarily of mind, or just as Democritus thought that it consisted of atoms, so the Pythagoreans thought that it consisted of numbers. To them mathematics was the whole of reality, and they did not distinguish between a geometrical solid and a physical body which would move in space.

They also studied numbers in a geometrical setting, paying

special attention to what they described as triangular and square numbers. The triangular numbers were 1, 3, 6, 10, 15, etc., because any one of these numbers of dots will exactly fill up the interior of a triangle with evenly spaced points, as in fig. 5.* The square numbers were similar, the triangle being replaced by a square; thus they were 1, 4, 9, 16, 25, etc. The Pythagoreans discovered a number of trivialities about these numbers, as, for instance, that the sum of two consecutive triangular numbers is a square number; this can be seen immediately on fitting one triangle on to the other, as in fig. 6.†

On the other hand, they made discoveries in true geometry which were of fundamental importance. The famous 'Theo-

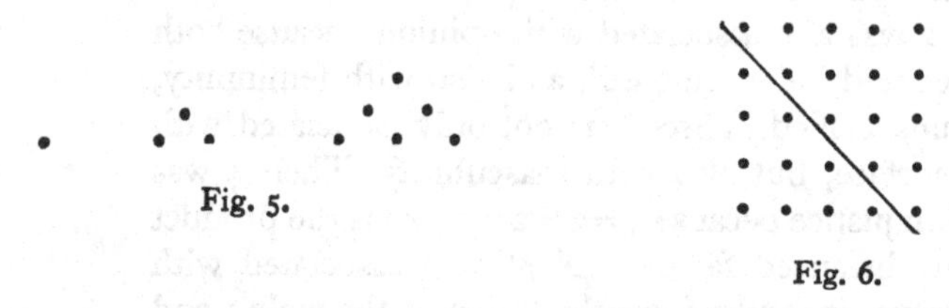

Fig. 5.

Fig. 6.

rem of Pythagoras' is usually attributed to them, (but see p. 10) and generally to Pythagoras himself: 'If a triangle is right-angled, the square on its longest side is equal in area to the sum of the squares on the other two sides.' Much of what the Pythagoreans did was silly, useless and misleading, but if they really discovered this theorem they laid a real corner-stone of mathematical science, lasting and indispensable. Pythagoras may have rated this as his greatest achievement, for Apollodorus, a poet of unknown date, writes of how 'Pythagoras discovered that famous proposition on the strength of which he offered a splendid sacrifice of oxen'. Yet such an action seems quite at variance with all we know

* Thus the nth triangular number was the sum of the series $1+2+3+\ldots$ to n terms, which is $\frac{1}{2}n(n+1)$.

† This is the geometrical equivalent of the relation

$$\tfrac{1}{2}(n-1)n+\tfrac{1}{2}n(n+1)=n^2.$$

as to the character of Pythagoras, and the story is so suspiciously like that already told of Thales (p. 22) that it seems possible that Apollodorus merely confused the two men, and was followed in his error by later writers. Even if the sacrifice was made by Pythagoras, there is some uncertainty as to which particular discovery produced it. Most accounts say it was the theorem just described, but one at least says it was a different theorem, while Vitruvius, one of the earliest writers on this subject, says that the sacrifice resulted from the simpler discovery that the particular triangle of sides proportional to 3, 4 and 5 is right-angled. The Pythagoreans might well have discovered this through their studies of 'square' numbers.*

It is often said that this last result was known to the ancient Egyptians, and that their 'rope-stretchers' used it to construct right-angles, but apparently there is no valid evidence for this.† On the other hand, as we have already seen, the general theorem was known to the Babylonians of about 1700 B.C. Tablets of this date discuss how to calculate the diameter *AE* of a circle (fig. 7) when the chord *BC* and the sagitta *DA* are known, the result obtained being simply an expression of the Pythagorean theorem

$$(OC^2 = OD^2 + DC^2).$$

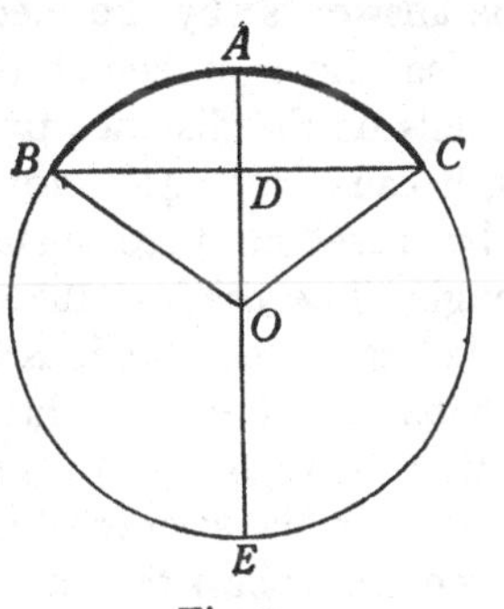

Fig. 7.

An Indian book of the fourth or fifth century B.C. also states

* For we may represent any square number n^2 by $n \times n$ dots arranged in a square, and can then add a fringe of $2n+1$ dots, n round each of two adjacent sides and one in the corner, thus obtaining the square number $(n+1)^2$. If, then, $2n+1$ is itself a square number, a^2, we have

$$a^2 + n^2 = (n+1)^2,$$

so that a, n and $n+1$ form possible sides for a Pythagorean triangle, a formula which is attributed to Pythagoras himself. The smallest value of $2n+1$ which is a square number is 9, and this leads to the triangle of sides 3, 4 and 5.

† T. Eric Peet, *The Rhind Mathematical Papyrus*, p. 32.

the general theorem, but without proof, and explains how right-angles can be drawn by constructing triangles of sides 3, 4, 5; 5, 12, 13; 8, 15, 17; and 12, 35, 37.

However much or little the theorem may have been known before the time of the Pythagoreans, there can be but little doubt that it was rediscovered independently by them—and, according to most of the European writers of the next few centuries, by Pythagoras himself. And it was this rediscovery that introduced the theorem into modern mathematics.

The reader with no mathematical interests may wonder wherein the importance of this theorem lies; he may think it abstract, pedantic and of only academic interest. But let us examine a practical application of it. Colchester lies 30 miles to the north and 40 miles to the east of London. How are we to deduce how far it is from Colchester to London in a bee-line? The answer is: by the theorem of Pythagoras, and by nothing else short of actual physical measurement. The theorem tells us the distance from Colchester to London is 50 miles, because $50^2=40^2+30^2$. Thus it tells us how much we save by travelling direct instead of going round the sides of a triangle. The original theorem was applicable only to right-angled triangles, but it is easily extended to triangles of any shape. Looked at in this way, we can no longer feel surprise that the theorem forms a corner-stone—perhaps *the* corner-stone—of the science of geometry.

We do not know how the result was proved. Hofmann has collected thirty different proofs, and it may have been almost any one of these. Perhaps it is most likely to have been the simplest, which runs as follows:

We drop a perpendicular AD from the right-angle A of the triangle on to the opposite side BC (fig. 8). Then the three triangles ABC, DBA and DAC are all of the same shape, so that their sides must be proportional (p. 21). Thus

$$\frac{BC}{AB}=\frac{AB}{BD}, \quad \text{whence} \quad AB^2=BC\times BD.$$

On treating the other small triangle in the same way, we find that

$$AC^2 = BC \times DC.$$

Thus the sum of squares on AB and AC is BC ($BD + DC$), which is equal to the square on BC—the theorem of Pythagoras.

We have not yet extracted all the meat from fig. 8, for the two small triangles are also similar to one another, so that $AD^2 = BD \times DC$. Here we have the solution of another problem which is attributed to Pythagoras—to construct a square which shall have the same area as a given rectangle. The 'splendid sacrifice of oxen' is sometimes associated with the discovery of this, rather than of the main Pythagorean theorem, but the two are so intimately connected that they may well have been discovered together.

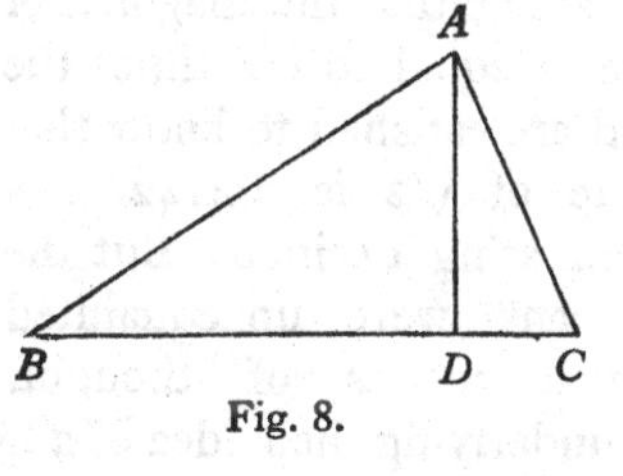

Fig. 8.

The Greeks were much interested in problems of this last type. They had little or no aptitude for algebra, so that even the simplest of algebraic formulae meant nothing to them unless they could draw a geometrical picture of its meaning. They knew that the areas of surfaces and the volumes of solids were important, but they did not know how to express them except as the areas of squares and the volumes of cubes.

The simple problem just mentioned may be described as that of squaring the rectangle. A far more famous problem was to square the circle, i.e. to draw a square of which the area should be equal to that of a given circle. It has long been known that this problem cannot be solved by purely geometrical methods, so that 'trying to square the circle' has almost passed into common language as a synonym for attempting the impossible. But this was not known to the Greeks, and the Pythagoreans were commonly credited with having solved the problem (see p. 37).

Another problem of a similar type, however, they did show to be impossible of solution. In any square $ABCD$ (fig. 9), the square on the diagonal AC is equal to twice the square on either of the sides AB, BC. This can be seen from the theorem of Pythagoras, or by completing the square on AC as in fig. 9—again a pattern such as might be found on a tiled floor. We express this by saying that the diagonal of a square is equal to $\sqrt{2}$ times the side, and are satisfied to know that the value of $\sqrt{2}$ is 1·4142..., a non-terminating decimal. But the Pythagoreans were unacquainted with such modes of thought.

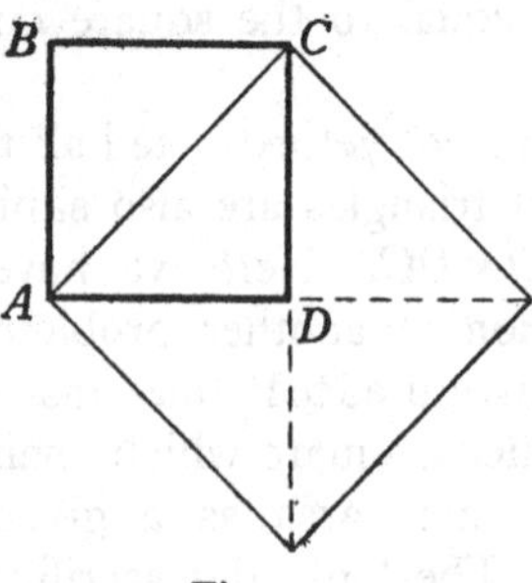

Fig. 9.

Always underlying their idea of a line was a picture of it as a sequence of minute units, all equal and so small as to be little more than points. If the side of any square contained p such units, and its diagonal q, then the number we denote by $\sqrt{2}$ would have the value q/p.

It is fairly easy to show that no such fraction can exist.*

We express this by saying that the square root of 2 is 'incommensurable'. The Pythagoreans seem to have discovered this incommensurability at an early date,† and realised that it made havoc of their doctrines that every line consists of a chain of equal finite units, and that nature is dominated by integral numbers. They are said to have tried to hush up

* For if it can, let p and q be the smallest numbers of which it can be formed. Then $q^2=2p^2$. Since $2p^2$ is an even number, q^2 must be an even number, so that q itself must be an even number; let us replace it by $2r$. Substituting this value for q, the original relation becomes $p^2=2r^2$, which is of the same form as the original relation, but is formed with smaller numbers. Thus the original supposition that p, q were the smallest numbers has led to a contradiction. It follows that the relation $q^2=2p^2$ cannot be satisfied by any numbers at all.

† One of the later Pythagoreans, Theodorus of Cyrene, Plato's teacher in mathematics, is said to have proved that the square roots of 3, 5, 6, 7, 8, 10, 11, 12, 13, 14, 15 and 17 are also incommensurable.

their fatal discovery, but the truth could not be concealed for ever, and it has been thought that this explains why the Greeks banished the idea of numbers and exact measurement from their geometry.

The mathematician Zeno (*c.* 495–435 B.C.; not to be confused with Zeno of Citium, a philosopher of a later date) possibly devised his famous paradoxes with a view to teaching the same lesson, although there are different opinions as to this. The best known is the paradox of Achilles and the tortoise who are to run a race, the tortoise receiving, say, 1000 yards' start. Can Achilles ever catch the tortoise? Zeno shows that if the Pythagorean ideas as to length are sound, he never can.

For Achilles will soon have covered the 1000 yards which the tortoise took as handicap, while the tortoise has covered only 100. The race may now be supposed to start afresh, but with the tortoise's start reduced to 100 yards. In a second stage, Achilles covers this 100 yards, while the tortoise covers only 10. And so the race continues, stage after stage, and at each stage the handicap is reduced to a tenth of its previous value. But this can never reduce it to zero, and after an infinite number of stages, the tortoise will still be in front. Achilles and the tortoise have both covered an infinite number of stages, each consisting, according to Pythagorean conceptions, of a finite number of finite units, so that the total distance covered must be infinite. If the Pythagoreans were right, Achilles could never catch up the tortoise, all of which is, of course, absurd.

Finally, the Pythagoreans gave much attention to 'regular solids'—solid figures in which all the sides and all the angles are equal. They knew of four such solids which could be formed out of squares and equilateral triangles as their faces. The simplest is the cube, built up of six squares standing mutually at right angles. Then come the four-sided pyramid, or tetrahedron, built of four equilateral triangles; the eight-sided octahedron built of eight equilateral triangles; and a more complicated figure, the icosahedron, which had twenty equilateral triangles forming its faces. Finally, Hippasus discovered the dodecahedron, formed of twelve regular pentagons, in about 470 B.C. (p. 26). This, as we now know, completes the list of regular solids.

To the modern mathematician these various studies seem to be concerned only with comparatively trivial side-issues. To the Greeks, imbued with the idea that the universe was fundamentally something of perfect regularity, they seemed to be of the greatest importance. We shall see below how they survived into a later age and figured in its efforts to discover the arrangement and workings of the planets.

ARCHYTAS. The work of these earlier Pythagoreans was continued and extended by later generations of the society, although new interests also claimed their attention. Among the later Pythagoreans, special mention should perhaps be made of Archytas (about 400 B.C.), a most worthy person who was seven times governor of the city of Tarentum. He was especially interested in the mechanical applications of science, and is said to have worked out the theory of the pulley. He also constructed a number of mechanical toys, including flying birds, so that perhaps we ought to regard him as the father of the science of aeronautics. This broadening of Pythagorean interests did not commend itself to all the members of the society, and when Archytas was finally drowned in a shipwreck, some of the more conservative members of the society averred that this was a very suitable ending for one who deviated so far from the lines of study laid down by their founder.

Archytas became famous for his solution of the problem of 'duplicating the cube'—a far more difficult problem than the duplication of the square already mentioned. It had been one of the famous unsolved problems of antiquity, known as the Delian problem for the following reason.

In or about 430 B.C. (so Philoponus tells us), the Athenians were afflicted with an outbreak of pestilence (probably typhoid fever), and sent emissaries to the Temple of Apollo at Delos to inquire how it could be stopped. The oracle told them to double the size of the altar of Apollo in Athens, which was cubical in shape. On receiving this advice, the Athenians doubled the height, length and breadth of the altar. They expected the pestilence to stop, but it only got worse. When a second deputation was sent to Delos, it was explained to them that the three-fold doubling of the dimensions of the altar had not merely doubled the volume, as the god had demanded, but had increased it eight-fold. Hence the importance of knowing how to make a cube twice the volume of a given cube.

The solution which Archytas gave was very intricate, depending on the properties of the complicated curve in which a rotating semicircle cuts a stationary cylinder; it is of interest as showing to what a high degree of geometrical knowledge and skill the Pythagoreans of this time had attained. We should also notice the remarkable skill shown by the priests who served the Delian altar in setting a problem which they had probably hoped would not be solved before the pestilence had run its course.

THE ATHENIAN SCHOOL

While the Pythagorean school was declining in numbers and strength, a new scientific school was developing in Athens, which had now become the capital and cultural centre of Greece. To understand how this had come about, we must recall the history of the years 490–480 B.C.—the decade of Marathon, of Thermopylae and of Salamis.

To the east of Greece lay the kingdom of Persia, with its rapidly increasing power and growing ambitions. Its Emperor Darius wished to expand it to the west, and so came into conflict with the Ionian settlements fringing the coast of Asia Minor, as well as with the cities of Athens and Eretria which sent them help. Greece was not yet a united nation, but a collection of isolated city-states, each owning allegiance only to its own local government, so that after the Ionians had been defeated, Athens was left to face the impact of the Persians almost alone. Yet when the two forces met at Marathon in 490 B.C., it was the Persians who fled from the field.

Darius's successor Xerxes next attacked the still disunited Greeks with a vast army which was reputed, although no doubt erroneously, to be five million strong. The Spartans sent a small force to oppose it, and when this was annihilated, to the last man, at the vital pass of Thermopylae, the whole of Attica, including Athens, lay open to the enemy. There were clashes at sea off Salamis, and on land at Plataea. When the enemy were utterly defeated in both, they retired from Europe and the menace from the east was, for the time being, removed. But with a view to avoiding similar troubles in the future, the various city-states welded themselves, by the confederation of Delos, into a single nation with Athens as their capital.

HIPPOCRATES OF CHIOS. In this fifth-century Athens, we find three mathematicians of importance. First comes Hippocrates of Chios (not to be confused with the more famous Hippocrates of Cos, the physician), who was born in Chios, one of the Ionian islands, in about 470 B.C., and was reputed to have squared the circle. He started life as a merchant, and it is said that he went to Athens at the age of about 40 to safeguard his interests in a law-suit, took to consorting with the teachers and philosophers there, and finally opened a school of his own.

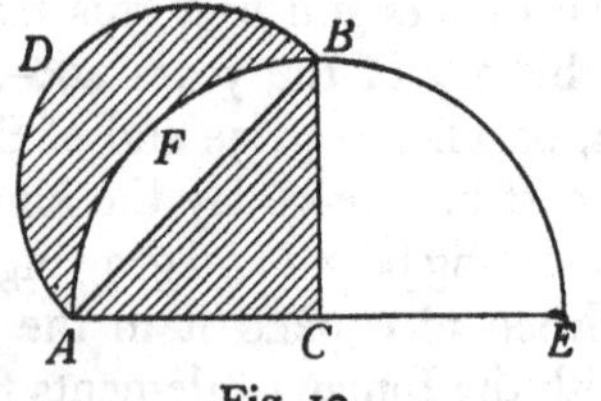

Fig. 10.

He had not, however, squared the circle. In fig. 10 the line *AB* is the base of the right-angled triangle *ACB*, and is also the diameter of the circle *ADB*, so that the quadrant *ACB* of the larger circle will be equal in area to the semicircle *ADB* of the smaller circle. Subtracting from each of these areas the part *AFB* (unshaded) which they have in common, we find that the shaded 'lune' *ADBF* is equal in area to the shaded triangle *ACB*. Thus Hippocrates had squared the lune *ABFD*, which was rather like a circle, at least in having curved boundaries—hence his reputation.

Hippocrates also tried, but unsuccessfully, to duplicate the cube. Indeed, the Athenian mathematicians concentrated especially on three problems:

(1) The duplication of the cube,
(2) The squaring of the circle,
(3) The trisection of the angle,

all of which look simple, but are now known to be impossible of solution by the ruler-and-compass methods by which they tried to find solutions. The explanation of this apparent piece of bad luck was perhaps that all the simple-looking problems had already been solved, except the few which were insoluble.

And so it came about that the best work of Hippocrates was nothing more than a text-book on geometry—the earliest of which we know and, according to Proclus, the first ever written. We know little of its contents, but it may have had some influence on, and perhaps even served as a model for, the more famous *Elements of Geometry* of Euclid, which appeared some years later, in which case Hippocrates must in some part have been responsible for the way in which geometry was taught in the schools of Europe for more than 2000 years.

PLATO. We come next to Plato, the great philosopher. He was born in Athens in 429 B.C., and became a pupil of Socrates in 407. He took to travel when the Athenians put Socrates to death in 399 B.C., and studied mathematics in many countries. He returned to Athens about 380 B.C., and founded the school known as the 'Academy', which lasted for nearly a thousand years. He died in 347 B.C.

Plato's fame of course rests on his philosophy, but his writings show that he had a good knowledge and understanding of mathematics. Yet only one mathematical result of any consequence has ever been attributed to him, and this perhaps erroneously. It is another attempt at the duplication of the cube.

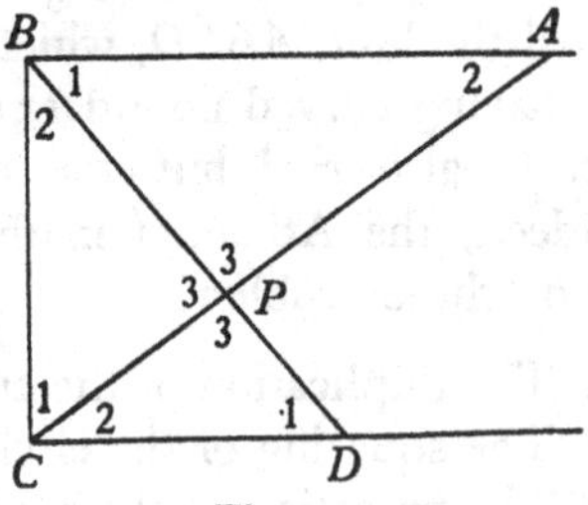

In fig. 11, the angles at B, C and P are all right angles, so that the three angles marked 1 are all equal, as are also the three angles marked 2, and the three angles marked 3. Thus the three triangles which have P as their vertex, namely, APB, BPC and CPD, are all similar, whence it follows that

Fig. 11.

$$\frac{PA}{PB}=\frac{PB}{PC}=\frac{PC}{PD}, \quad \text{so that} \quad \left(\frac{PA}{PB}\right)^3=\frac{PA}{PB}\times\frac{PB}{PC}\times\frac{PC}{PD}=\frac{PA}{PD}.$$

If, then, we can arrange for PA to be twice PD, a cube of edge PA will have just double the volume of a cube of edge PB, and the problem of duplicating the cube is solved. There is no way of arranging this with ruler and compasses, but it is easy to do it with a mechanical arrangement in which rods slide over the surface of a plane board, while pins in the rods slide in grooves cut in the board. Plato disapproved of any instruments being used beyond ruler and compasses, yet the use of the instrument just described might have saved Athens from the pestilence!

Although Plato's direct contribution to mathematics was small or nil, he must have exercised an immense influence on the development of the subject. He insisted that it should be taught in its abstract aspects, and not for utilitarian ends. The Pythagoreans had professed similar ideals, one of their maxims being 'a theorem and a step forward, not a theorem and sixpence'. Studied in this spirit, Plato considered it a model for all other studies, because of its certitude and exactness, and thought it the best training in logical thought. Over the entrance to his Academy was inscribed ἀγεωμέτρητος μηδεὶς εἰσίτω (no one may enter who is not a mathematician)—no empty threat, for we are told of a candidate being rejected because he knew no geometry. Indeed, there seems to have been no limit to Plato's faith in the educational value of mathematics. When he was over sixty years of age, Dionysius II, the young Tyrant of Syracuse, summoned him to instil wisdom and virtue into his court. Plato tried to do this by teaching them all geometry until, in the words of Plutarch who tells the story, the whole Palace became 'one whirl of dust' as the Prince and courtiers drew their diagrams on the sanded floors. But the Prince soon decided that other methods led more directly to the desired result, and Plato returned to Athens.

EUDOXUS. A third member of the Athenian school who should be mentioned here is Eudoxus (408–353 B.C.). His astronomy was more important than his mathematics (p. 66), although the latter, unhappily all lost, appears to have been of first-rate quality.

MENAECHMUS AND THE CONIC SECTIONS. Eudoxus left Athens to found a school in Cyzicus, and here his pupil Menaechmus (375–325 B.C.) initiated the study of the conic sections.

If we cut a solid body through with a knife or a saw, or if we imagine a geometrical solid being intersected by a geometrical plane, we obtain a 'cross-section' of the solid, which will be bounded by a curve which we may call the 'curve of cross-section'. For instance, the curve of cross-section of a cricket-ball is always a circle, no matter how the section is cut. More complicated solids naturally give rise to more complicated cross-sections. For instance, the curve of cross-section of a cylinder or a cucumber is a circle if the section is made at right angles to the axis, but otherwise is the curve we call an ellipse—a sort of elongated circle.

Menaechmus examined the curve of cross-section of a cone, and found that it could be any one of the three curves we now describe as conic sections, namely ellipses, parabolas and hyperbolas (fig. 12).

The conic sections thus introduced into science were destined to play a great part in the future growth of knowledge (p. 83), but their time was not yet. The immediate results which Menaechmus obtained were neither very important in themselves nor were they used for any important purpose; indeed, Menaechmus used them mainly to construct two more solutions of the now threadbare problem of the duplication of the cube.

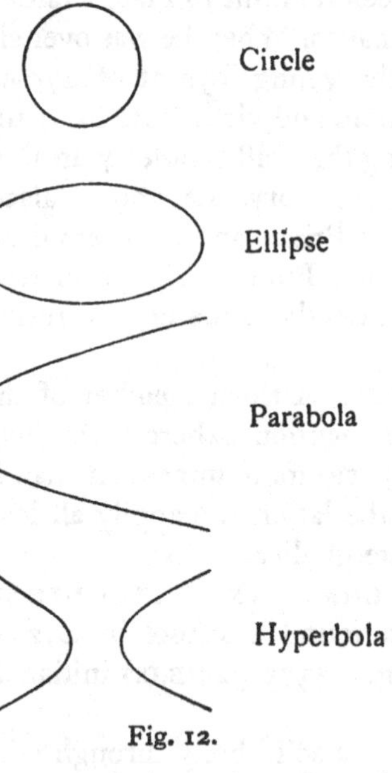

Fig. 12.

Not only this particular problem, but also the whole of Greek mathematics, was becoming rather threadbare by now. If we wish to trace out the more important steps in the progress of mathematics, and of scientific knowledge in general, we must turn our backs on Greece and move to Egypt, more particularly to the new and magnificent city of Alexandria, but before we embark on a study of Alexandrian mathematics, let us examine what progress physics and astronomy had made in the period we are now reviewing.

GREEK PHYSICS AND PHILOSOPHY

The modern physicist attacks his problems after a quite definite plan which was vaguely known to the Alexandrians (p. 124) and, as we shall see below, was pointed out to the moderns by Roger Bacon, Leonardo da Vinci, Francis Bacon,

Galileo and others. Its essence is to attack the problem not as a whole but piecemeal, and to start not from pre-conceived general principles but from firmly established experimental knowledge. Some special phenomenon, or some special property of matter, is singled out for detailed experimental study, in the hope that in this way law and order may be detected in one small corner of the universe. When this has been achieved, the field of knowledge is extended bit by bit, questions being asked of nature by direct experiment at every step.

Greek physics was something entirely different. The Greeks could not have followed the method just described, even if they had wanted to, from lack of experimental skill and equipment. But they would not in any case have wanted to, and this for two reasons.

In the first place, such a procedure was quite foreign to their modes of thought. They did not want scraps of knowledge about isolated corners of the universe, but a balanced and comprehensive view of the whole.

In the second place, their general attitude towards life resulted in many cases in a positive aversion against increasing knowledge by experiment. In the ordinary affairs of life, they esteemed mental activity far more highly than physical, which they thought unworthy of freemen and fit only for slaves. In some cities the freemen were not allowed to engage in mechanical trades. As Xenophon said: 'The mechanical arts carry a social stigma, and are rightly dishonoured in our cities. For these arts damage the bodies of those who work at them ...by compelling them to a sedentary life and to an indoor life, and, in some cases, to spend the whole day by the fire. This physical degeneration results also in the degeneration of the soul.' Experimental science naturally came under the shadow of this disapproval.* This attitude reached its culmination in Plato. Many before him had commented on the untrustworthiness of the human senses, but he went to the

* Farrington, *Greek Science*, p. 23.

length of arguing that their evidence should be used only to suggest ideal problems for discussion or intellectual gymnastics. 'While we live we shall be nearest to knowledge when we avoid, so far as possible, intercourse and communion with the body, and keep ourselves pure from it until God himself sets us free.' In astronomy he thought that the motions of the heavenly bodies should be studied only as providing approximations to ideal motions of absolute swiftness and absolute slowness; these absolute motions are to be apprehended only by reason and intelligence, and not by observation; 'as in geometry, we should employ problems and leave the heavens alone, if we would approach the subject in the right way'. He made a similar complaint about the musicians, who set their ears before their understanding: 'The teachers of harmony compare only the sounds and consonances which are heard, and their labour, like that of the astronomers, is in vain.'

It is not, then, surprising that the earliest Greek physics consisted mainly of abstract thought of a kind which we should now describe as baseless speculation. Making no contact with the outer world, and guided only by their individual ideas as to the fitness of things, the Greeks tried to discover the plan of the world out of their inner consciousnesses. Some assumed that the world must have been constructed by its maker after some simple and elegant pattern. Others, assuming that the circle was the perfect curve, concluded that most natural motions must take place in circles. Yet others supposed that there must be a sort of moral governance in the universe; for instance, Anaximander the evolutionist thought that all existing things must pass away in time, so as to make amends for the injustice they had committed in pushing their predecessors out of existence.

Out of all the resulting tangle of confused and inconsistent speculations, two main schools of thought stand out with some clearness. The visible universe consists of matter in activity; one school concentrated on the matter and tried to guess its nature, the other school concentrated on the activity and tried to guess its significance. One was concerned with the actors, the other with the play; very roughly they corresponded to our modern materialists and idealists. The first school drew its strength principally from Ionia, the second from Italy, consisting mainly of the Pythagoreans and their followers.

Ionian Materialism

It was natural that the Greeks, with their continual striving after elegant generalities, should conjecture that all the varied richness of nature could be explained by some simple formula. They began by attributing it to some common substance out of which the whole world was made. In the earliest epoch of European science, the question which loomed largest in the speculations of the philosophers was 'What are all things made of?'

Thales answered with the single word 'Water', but obviously this answer did not mean the same to him as it would to us. He said that 'that which exists' can take the three forms of mist, water and earth, by which he probably meant much what we mean when we say that matter can exist in the gaseous, liquid and solid states. He chose water as the fundamental form because he took the outstanding characteristic of the world to be its fluidity or wateriness; it was for ever changing in ways which would be impossible for a solid structure.

Anaximander the evolutionist took a different view, maintaining that the first principle and basic constituent of all things was 'a continuous infinite medium' which filled all space. It was the first appearance of the so-called 'ether', which remained in science until the present century. Anaximander's description of the physical functions of his ether reminds us of the similar descriptions given by the Victorian physicists, while his description of its philosophical functions—'out of it everything is generated, and into it everything returns'—reminds us of the description of space-time (p. 295) given by the twentieth-century philosopher Alexander.

Fifty years after Thales, his direct successor Anaximenes (*c.* 585–525 B.C.) compared things in their essential nature to air instead of water. The basic substance of the universe, he said was, πνεῦμα—breath, like the air we breathe; he held that, just as ordinary air sustains our human life, so a more general form of air sustains all the life of the universe. And for Anaximenes, as for Thales before him, all things were endowed with life. He further believed that the various forms of matter changed into one another through processes of condensation and rarefaction. Thus when water was rarefied, it became air, and when it was both rarefied and heated, it became fire, which was thus nothing but heated air. He also believed that the condensation of water produced earth, a belief of which traces survived until the seventeenth century (p. 262). Thus for Anaximenes the four elements—earth, water, air and fire—which were later to figure so largely in Greek speculative physics (p. 65), were all modifications of one another.

Another 50 years passed, and Heraclitus of Ephesus (*c.* 540–475 B.C.) taught that fire, the most changing of all substances, was the prototype of all things. Everything, he said, begins as fire, but fire changes into water and water into earth. His main doctrine was that everything is in a state of flux—πάντα ῥεῖ, καὶ οὐδὲν μένει (everything flows and nothing stands still); we never step into the same river twice.

ATOMISM. An entirely different set of doctrines was taught by the next Ionians of note, Leucippus of Miletus (of uncertain date), and his pupil Democritus (*c.* 470–400* B.C.), who quite possibly may also have been a Milesian. They held that the universe consisted of nothing but unchanging atoms and the space between them. The atoms were not only indivisible, as their name implied (ἀ-τέμνειν = not to be cut),

* But some accounts say that Democritus lived to be 90, 100, 104, 108 or 109 years old.

but were also uniform, solid, hard and incompressible. Their substance was indestructible, and so also was their motion; an atom continued to move so long as nothing checked it.

There was nothing vital in this picture; the changes in the universe did not result from intrinsic changes of the atoms, but from their motions and rearrangements, which happened from a compelling necessity. Thus the universe became a machine following a predestined path.

These doctrines took the emphasis off human perceptions and emotions, which now became unimportant incidents in the world, and stressed the existence of an objective world external to man, independent of man, indifferent to man. In brief, external nature had been discovered. The world which hitherto had been man's playground and pleasure ground became his prison. Hitherto it had been permeated by beauty, sweetness and warmth—the gifts of the gods to men—but these were no longer part of nature; they were imaginings of man himself. Democritus wrote: 'According to convention there is a sweet and a bitter, a hot and a cold; according to convention there is colour. But in reality there are atoms and the void. The objects of sense are supposed to be real, and are usually regarded as such, but in truth they are not. Only the atoms and the void are real.'

Physically these doctrines had much in common with modern atomic theory, but they were based neither on knowledge nor observation. Philosophically they were almost identical with present-day philosophical materialism and, like this, they implied a negation of free-will. Man could not choose what he would do; this had been decided for him long ago by the arrangement of his atoms. Determinism had entered science, but the Greeks called it 'compulsion'—ἀνάγκη φύσιος, the necessity of becoming.

The Pythagorean 'Elements'

While the Ionians were picturing the universe as something fundamentally simple, the Pythagoreans and the Sicilian Empedocles of Agrigentum (*c.* 500–440 B.C.) were advocating a more complex view of the world, replacing the one fundamental substance of the Ionians by four distinct 'elements' —earth, water, air and fire. Empedocles taught that everything was formed of these four elements mixed in different proportions under the influence of attractive and repulsive tendencies, and as his thought made no clear distinction between man and the inanimate world surrounding him, the same was supposed to be true of man, the attractive and repulsive forces now assuming the forms of love and hate.

The four elements were themselves formed by the attractions and repulsions of two pairs of contrasted qualities—hot and cold, wet and dry. Thus there were combinations according to the following scheme:

	Dry	Wet
Cold	Earth	Water
Hot	Fire	Air

We shall see how these ideas, remote from the truth though they were, were destined to play no small part in the later developments of physical thought.

Empedocles taught that the universe had begun as a chaotic mixture of the four elements. First air was separated out of the mixture, and then fire; these were followed by earth, from which water was squeezed out. The heavens were formed from the air, and the sun from the fire, while the 'other things' about the earth were formed of the remainder.

Empedocles made a more valuable contribution when he taught that light travels through space at a finite speed; it takes time to pass from one place to another, from the object seen to the eye that sees.

PLATO AND ARISTOTLE AS PHYSICISTS

While physics was still in this primitive stage of its development, it met with two major disasters in the attitudes of two great thinkers, Plato and Aristotle. Plato was unsympathetic and even contemptuous, while Aristotle failed to understand the function which physics ought to fulfil.

PLATO. We have seen that Plato was a mathematician of no mean powers, who consistently professed a high regard for mathematical studies. But this was because mathematics dealt with things mental, and not because it opened the road to a better understanding of things material; his admiration was for pure mathematics, not for what we now call applied mathematics, which did not exist in his day. Like the philosophers of many ages, he saw that our only certain knowledge is of the sensations that affect our minds. These may seem to originate in an outer world of matter, but the existence of such an outer world is only a hypothesis. Mind may be the only reality, and the outer world only an invention of our minds. Plato accepted a variant of this latter view; he saw mind as the only fundamental reality, and the material world only as a shadow of reality. We had come into the world, he maintained, with a number of general ideas inborn in our minds, such as the ideas of hardness, redness and sphericity. These ideas he called 'forms'. When we say we see a hard red ball, we merely mean—so Plato said—that something which is affecting our senses seems to fit into the forms, already in our minds, of hardness, redness and sphericity. The object may fit the forms well or badly, but in no case will the fit be perfect; no material object can be quite so perfectly spherical as our mental idea of a sphere, or so completely red as our idea of redness. Plato, believing that perfection and reality necessarily went together, argued that the eternal and unchanging forms must be the true realities of the world, while the material objects which come and go, and at best provide only fleeting impressions and imperfect representations of the forms, have a lower degree of reality; their relation to the realities is that of the circles which the mathematicians draw in the sand to true circles. (We can see why Plato thought that physical problems should be discussed as idealised by our minds rather than as presented by our senses). Thus in its ultimate essence the world resembles neither water nor air nor fire nor hard atoms, but mind.

A modern scientist would challenge all this on the ground that the Platonic forms are not inborn in our minds, but are classifications which our minds create out of experience. He would say, for instance, that a blind man cannot have the form of redness in his mind, nor a deaf man the form of trumpet-tone; if we find these forms in our own minds, it is because we are not blind and deaf, but have seen and heard much in our lives. But Plato, obsessed by his doctrine of inborn forms, found the study of the shadowy substance we call matter unimportant. More than most men he had a vision, true or untrue, that the only thing worth while for humanity was the search for the good and the beautiful—two qualities which the Greeks identified so completely that they used the same word for both. Thus he specially hated the doctrines of Democritus which explained humanity, goodness and beauty as mechanical manifestations of material atoms. He never mentions Democritus by name, but is said to have expressed a wish that all his books might be burned.

We can see his general attitude to physical science from some remarks which he puts into the mouth of Socrates.* The astronomer Anaxagoras (p. 60) had written a book which first asserted that 'In the beginning all things were mixed up, then mind came and reduced them to order', and then proceeded to explain in mechanical terms how this had been done. Socrates says that he had expected that the book would first tell him whether, for instance, the earth was round or flat, and then would go on to explain the reason for it, namely, that 'it is better that the earth should be as it is' than that it

* *Phaedo.*

should be any other way. He goes on: 'For I could not imagine that, when he had once said that these things were ordered by mind, he would have assigned to them any further cause except the fact that it is best that they should be as they are. These expectations I would not have sold for a large sum of money. From what a height of hope, then, was I hurled down when I went on with my reading, and found my philosopher forsaking mind and any other principle of order, and having recourse to airs, ethers, waters and other eccentricities.'

There could hardly have been a more complete misunderstanding of the aims and methods of physics.

ARISTOTLE. Plato's attitude was disaster number one for physics, but worse was to come from his pupil Aristotle. At the early age of 17, Aristotle had left his birthplace Stagira in the crude, semi-barbaric state of Macedonia, north of Greece, where his father was court physician, to study with Plato in Athens, and remained there until the death of his teacher 20 years later. He then lived in Lesbos, one of the islands off the coast of Asia Minor, for the five years 347–342 B.C. After this he spent six years as tutor to the young Prince Alexander of Macedonia who was later to be known as Alexander the Great, the conqueror of the greater part of the civilised world, and the founder of an empire which extended from Greece to India and from Thrace to Egypt. In 334 B.C. he returned to Athens where he became a public teacher and founded the famous school of the Peripatetics. Here he does not appear to have met with the unqualified approval of his brother philosophers, many of whom objected that his manners were suited to a Court rather than to an Academy; a long dirty beard and shabby clothes were not for him. In 323 he again left Athens, and died the year after.

In his youth he had been noted for the voracity of his reading; in mature years he acquired an encyclopaedic mind which took all knowledge for its province, and so invaded every branch of science. He wrote on a vast variety of subjects, pouring out torrents of clear thought and good sense, which were generally controlled by a penetrating judgement and a profound store of knowledge. But in science his attainments were very uneven; he was a forcible biologist, but a feeble physicist. His biology was based on personal observation, and some competent judges consider him to have been one of the greatest biological observers of all time. Some of his observations retained their importance right down to modern times, and his classification of the forms of life was not superseded until the time of Linnaeus. But his acute powers of observation led him nowhere in physics, since the physical world is too complex for its secrets to be unravelled by a mere inspection. Planned experiments are needed here, and the idea of experimentation was utterly alien to the outlook both of Aristotle and of his contemporaries.

In every experiment we assume that an event is an effect which is preceded by a cause; we provide the cause and observe the effect, thus studying one link in the cause-effect chain which we believe to run throughout nature and to govern its happenings. This cause-effect chain did not enter into Aristotle's thoughts. To the question 'Why is so-and-so (*A*) the case now?' our modern answer is of the form 'Because so-and-so (*B*) was the case in the past'. But Aristotle's reply was of the form: 'Because it is in the nature of *A* to be as it is.' For instance, we answer the question: 'Why is the moon eclipsed?' by saying that the earth has moved between the sun and the moon. But Aristotle regarded it as a sufficient answer to say, 'Because it is in the moon's nature to be eclipsed'. 'It is clear', he wrote, 'that the nature of the thing and the reason of the fact are identical.'*

Such a mentality obviously disqualified Aristotle as a physicist. It has been suggested that he had done his biological work while living in Lesbos between the ages of 37 and 42, but produced his writings on physics after his return to Athens. He was then already in the fifties, so that his mind

* See F. M. Cornford, Essay on 'Greek Natural Philosophy and Modern Science' in *The Background to Modern Science*, p. 11.

may no longer have been impressionable to new ideas, but became set in the mould of his biological thought; all science may have seemed to him a matter only of observation and description.

Apart from this, Aristotle still held an entirely homocentric view of the world, seeing man as the centre of all creation, its climax and final triumph. For him the universe was primarily a universe of human sensations, and the ultimate truth about anything was told by the sensations that it produced in the human body; the most that we could ever know about honey was that it was brown and sticky, wet and sweet. Aristotle regarded these qualities as absolute, not as relative to the mind that perceived them. He harnessed these bad philosophical ideas to a bad system of mechanics. Democritus had taught that a moving body continued in its motion until something intervened to check it, but Aristotle interpreted all motion as a gratification of natural inclinations—as though all things were living organisms. Just as a seed wanted to germinate and push its way up through the soil, so a body which was heavy wanted to sink, and one which was light wanted to rise; everything strove to reach its 'natural place' in the world. Thus smoke rose, and a stone fell. Convinced from the biological analogy that all things must either attract or repel one another, Aristotle accepted the four elements of Empedocles—earth, water, air and fire—as constituents of matter, but added a fifth, the 'quintessence', to form the basic substance of the universe. To justify this, he argued that two kinds of motion are possible, up-and-down and circular. Of the four elements of Empedocles, air and fire move up while earth and water move down. There must obviously, then, be a fifth to move in circular motion, and this can only be the ether of which the stars are made—an ether which is more divine than the other four elements, and must also be changeless, since there is no record of any change having ever occurred in the outer heaven or in any of its parts.*

* Sir T. L. Heath, *Greek Astronomy*, p. xlvii.

Aristotle is generally credited with the invention of formal logic—the logic of rigorous proof—and some think this was an even greater disaster to science than his physics. He was right in insisting that no fact could be certain unless it had been deduced by strict logic from other facts which were certainly true, but he failed to see that this is just what we can never do in science. In requiring all sciences to have the certainty of mathematics, Aristotle imposed on them the limitations of mathematics, which can never gain new knowledge, but only transforms old, and presents it in a new dress. He wrote voluminously on a great range of physical questions, but his method was always that of deduction, and as his premisses were almost invariably wrong, his conclusions were so likewise. Nearly 2000 years were to pass before the deductive methods of Aristotle were discarded in favour of inductive methods (p. 124), and then progress became rapid indeed.

In the meantime, the dead hand of Aristotle lay heavy on physics. Had it been otherwise, free discussion and a blending of the ideas of Democritus and Empedocles—atoms and forces—might have given physics a good start, for it is surprising how many of the basic ideas of modern physics were foreshadowed in the speculations of these two men.

The Epicureans and Stoics

The period which followed the death of Aristotle was one of general confusion and ferment—military, political and intellectual. Philip had conquered Greece, and the Greeks, smarting under their military defeats, had lost their former joyous self-assurance and irresponsible cheerfulness, and were feeling the need for a philosophy or religion which would instruct them how to live. Their easy-going Olympian religion had never done this; Christianity would bring its own solution in time, but the time was not yet.

Into this disturbed society were born the two new philosophical systems of Epicureanism and Stoicism. Both were appropriate to the grimness of the times. Epicureanism was a philosophy of contentment and happiness even under misfortune, while stoicism was one of self-control and devotion to duty. Both were primarily systems of ethics and religion, but as both invaded the realm of science, they are of some interest in the present book.

The Epicureans. Epicurus, the founder of the former system, was born in Samos, one of the Ionian islands, in 341 B.C., to a life of feeble health and grinding hardship which no doubt called for a new philosophy to make it tolerable; and so he taught the pursuit of the simple life, of mental calm, and of inward quiet. He was no scientist, despising knowledge in itself, and particularly 'the vanities of astronomy'. He dismissed the calculations of Aristarchus on the sizes of the sun and moon (p. 87) with the remark that the sun was probably about as large as it looks, or perhaps smaller, since distant fires often look larger than they actually are. Indeed he accepted an estimate of Heraclitus (p. 44) that the sun was about a foot in diameter; this would place the sun at a distance of only 115 feet from our eyes.

He taught a wholly materialistic physics, which denied to mind the position which Plato and Aristotle had assigned to it as the fundamental ingredient of the universe. He had studied the works of Democritus, and claimed that all existence is corporeal (τὸ πᾶν ἔστι σῶμα); there can be nothing but atoms and the void. There must be a void, or the atoms would have no room to move about, and it must be infinite in extent for it could only be bounded by something else of a different nature, and this cannot exist if the atoms and the void form the sum total of all existence. The atoms must be infinite in number, otherwise they would drift and scatter through the infinite void, not being kept in place by their collisions with other atoms. The atoms dart about through the void with incredible speeds, 'swift as thought', their rearrangements continually giving rise to new worlds. Objects for ever emit thin filmy images of themselves from their surfaces, and these, travelling in all directions, give rise to sensations when they impinge on our bodies; these sensations give us our knowledge of the world.

Epicurus was especially concerned to discredit the idea of a divine governance of the world. The gods, he said, are finer and higher than we are, but are just as much the products of nature as ourselves, and so are equally subject to natural laws. They cannot, then, govern the world, and are indeed indifferent to human affairs; man can still be master of his fate and captain of his soul. In this way Epicurus tried to set men free from what Lucretius described as the 'burden of religion',* although he conceded that they might, if they so wished, remain loyal to their traditional gods.

The Stoics. The other new philosophical system was founded by Zeno, a Phoenician by extraction, who was born in Citium in Cyprus, and came to Athens in 311 B.C., where he established a school and taught in the Stoa or painted Porch. His philosophy also was practical and suited to the needs of the times. It too taught a renunciation of the world; men were to be guided by their consciences and reason rather than by their desires, affections or emotions. The best minds and the noblest characters of the pagan world were either adherents of Stoicism or came directly under its influence.

Like Epicureanism, it taught a completely material system of physics; even such things as virtues and activities are described as bodies (σώματα). Every body was supposed to consist of an active and a passive principle, these being respectively the inert matter which could undergo change,

* Lucretius, *De rerum Naturae*, I, 63.

'When Man's life upon earth in base dismay,
Crushed by the burthen of Religion, lay,
Whose face, from all the regions of the sky,
Hung, glaring hate upon mortality,
First one Greek man against her dared to raise
His eyes, against her strive through all his days...
Till underfoot is tamed Religion trod,
And, by his victory, Man ascends to God.'

(Translation by Gilbert Murray, *The Five Stages of Greek Religion*, pp. 134–5).

and the force causing this change. Thus there was a reason for every change, and so for every happening and motion in nature. Two thousand years before Newton, the Stoics introduced the idea that every event occurred in accordance with universal law. The stars which, it was now being discovered, moved according to perfectly regular laws, must form part of a majestic and purposeful plan. Thus the world must be moving towards a perfection which had been designed by God, but could be achieved in part by man, so that human life became a thing of dignity and value.

These physical theories of the Stoics and Epicureans had but little effect on the development of scientific thought in Greece.

For the reputation of Aristotle stood so high that what he had said was deemed unchallengeable; if Aristotle had spoken, it was so, not only in the Greek world in which he lived, but also in the medieval world which was to come. Here the Christian Church supported his doctrines, which certainly fitted the spirit of religion better than the materialism of Epicurus and Zeno, and physics became crystallised in an Aristotelian mould until men began to think for themselves at the time of the Renaissance, until Stevinus and Galileo began to experiment to discover whether things were as Aristotle had said, and found they were not.

The Growth of Experiment

Such were the main lines followed by Greek physical thought, but it would be a mistake to think they were the only lines along which thought tried to progress. The idea that the nature of the universe could be unravelled by pure intellect, and without any appeal to facts, carries its own condemnation, and some of the Greeks must have been vaguely conscious of this. Thus in spite of their general aversion from appealing to facts, some of them did precisely this. Perhaps the first instance is to be found in an observation of Anaximenes (*c.* 550 B.C.) that if we blow gently on the back of our hand, the breath feels warm, whereas if we blow violently, it feels cold. Anaximenes interpreted these facts wrongly, but we recognise the true experimental method of appealing to nature for information, and noting her answer.

A few years later—possibly even earlier—the Pythagoreans were experimenting on the pitch of musical sounds. It must have been well known that deep-pitched sounds are produced by large structures, and high-pitched sounds by small; the lion roars, but the mouse can only squeak. The Pythagoreans tried to establish a relation between size and pitch. Boëthius, writing in the sixth century A.D., tell us that Pythagoras himself was passing a blacksmith's forge, when he was struck by the musical sequence of sounds given out by the hammers as they struck the anvil. He weighed the hammers, and found that the weights of four stood in the simple relation of 12:9:8:6,* while that of the fifth, which gave out a discordant note, stood in no simple relation to those of the others.

Pythagoras is said to have followed this discovery by a series of experiments on strings, and to have come upon certain laws which still form the basis of the science of acoustics. We know that strings which differ in length, but in nothing else, produce notes of different pitch. When two such notes are sounded together (or in succession, as the Greeks would have sounded them) the combination may be pleasant or unpleasant. The great discovery of the Pythagoreans was that it is pleasant only if the lengths of the strings stand in some quite simple numerical relation to one another, such as two to one, or three to two. It is more than 2000 years since they discovered the facts, but no one has yet found a completely satisfying explanation of them. We note them here as evidence of the increasing reliance on experiment; incidentally, they also establish acoustics as the oldest of the experimental sciences.

Nearly a century later, we find Empedocles (p. 46) in-

* If this was so, it was a pure coincidence, for there is no clear relation between the weights of hammers and the notes they produce from an anvil.

vestigating the nature of air by experiment. He placed the lower end of a tubular vessel in water, and proceeded as we do with a pipette or dropper. So long as his finger closed the upper end of the tube, the pressure of air inside kept the water out. If he took his finger away, the water could enter, and if he now again closed the end with his finger, the water could not get out of the tube even though this were lifted right out of the water; the pressure of the air outside the tube kept the water in. He interpreted this as showing that air was a substance, and was capable of exerting pressure. A few years later, Anaxagoras repeated the experiment, and also blew up bladders and showed that force was needed to compress them. The idea of the experimental appeal to nature was by now becoming familiar.

GREEK ASTRONOMY

Early Astronomical Pictures

Most races, even the most primitive, have invented stories for themselves to explain the general appearance of earth and sky, the alternation of day and night, and the simpler astronomical phenomena. Most, too, have devised cosmogonies to explain how things came to be as they are. The Greeks formed no exception, but their astronomy, like their mathematics, showed the influence of their scientific predecessors, the Babylonians and the Egyptians.

The Babylonians had pictured the universe as a vast room, with the sky as its ceiling and the earth as its floor. This floor was surrounded by water as a castle is by a moat, and on the far side of the moat lay mountains which supported the dome of the sky. More mountains covered with snow rose from the middle of the floor, and in these rose the river Euphrates, the centre of Babylonian life.

The Egyptians drew a similar picture, except that they put Egypt at the centre of the floor, possibly imagining that the yearly flooding of the Nile showed this to be the lowest part of the earth's surface. Four huge columns supported a ceiling from which stars were suspended like lamps.

The earliest Greeks adopted this general picture, but soon began to improve upon it. At the time of Homer (say the ninth century B.C.), they imagined the earth to be a flat disk, with the ocean, which they called the river Oceanus, replacing the surrounding moat. Above was the vault of heaven; below lay Tartarus, the abode of the dead, forming a second vault symmetrical with the vault of heaven.

THALES AND ANAXIMANDER. In the sixth century before Christ, Thales and Anaximander again amended this picture. Thales thought that the earth floated in water, while Anaximander detached it still further. Seeing that the stars revolved round the pole star, he concluded that they were attached to a complete sphere, the earth being suspended freely in space at its centre, without support of any kind. He imagined that the earth could stand in equilibrium in this way because it was at equal distances from the other heavenly bodies—almost as though he were already thinking of the earth as acted on by gravitational forces from the other masses of the universe.

This was an obvious improvement on the Babylonian, Egyptian and Homeric pictures of the universe, none of which could explain where or how the sun spent the night. It was now possible for the sun to pass under the earth at night, and not merely be carried round the encircling moat in a boat, as the Egyptians had imagined. But the picture was too revolutionary to command general assent; perhaps, too, it called for too much mathematical imagination. Shortly afterwards we find Anaximenes writing that the sun and the stars do not pass under the earth, but the sphere to which they are affixed turns round above the earth 'as a cap can be turned round on the head'—a conjecture which a minimum of observation would have instantly disproved.

Anaximander did not make his earth a sphere, but a circular disk or stumpy cylinder of thickness only a third of its diameter. He said that the sun was of the same size as the earth, and that it moved round the earth in an orbit which was 27 or 28 times as big as the earth, while the orbit of the moon was 19 times as big as the earth. But neither reasoning nor observation lay behind all this, only baseless conjecture. Indeed, the Greek powers of observation were still lamentably low. No one seems yet to have noticed that the bright part of the moon's surface always faces the sun, or to have conjectured that the moon owes its illumination to the sun. Instead, we find Anaximander saying that the stars in general, including the moon, had pipe-like passages projecting from them, through which we see their light. The waxing and waning of the moon resulted from such a passage being alternately opened and closed; if it was altogether closed, then an eclipse occurred.*

THE PYTHAGOREANS. In the century following Anaximander, many important advances in astronomy were made by Pythagoreans, although as usual it is difficult to assign ideas to their individual authors.

In place of Anaximander's stumpy cylinder suspended freely in space, Pythagoras is said to have believed that the earth was spherical in shape as, also, were the sun, moon and stars. The Pythagoreans also made the further great advance of supposing that the earth did not form a fixed centre to the universe, but revolved, with all the other planets, round a central fire; one account attributes the idea to Philolaus (p. 88), another to a certain Hicetas of Syracuse. If they had taken the further step of identifying this central fire with the sun, they would have made one of the greatest advances in the whole history of science. But for some reason or other, they never took this step. It has been suggested that the central fire was really meant to be the sun, but that they did not dare to say so from fear of encountering religious persecution such as afterwards befell Anaxagoras (p. 63). On the other hand, Aristotle says that they were convinced that the

* Hippolytus, *Refutatio omnium Haeresium*

total number of moving objects in the sky must total up to their mystical number ten. Towards this total the sun, moon and five known planets contributed seven, and the earth and the sphere of the fixed stars two more. To bring the tally up to the necessary ten, they imagined a 'counter-earth' which also revolved round the central fire. As they never saw either the central fire or the counter-earth, they had to suppose that the hemisphere of the earth on which they lived was perpetually turned away from both. This is obviously inconsistent with the central fire being the sun.

The hypotheses of the central fire and the counter-earth soon became untenable. Navigators were already beginning to sail out of the Mediterranean to explore the coasts of Africa to the south and of Europe to the north; soon they would voyage round the coasts of Britain to the frozen seas beyond. They saw many strange sights, but none to suggest the existence of either a central fire or a counter-earth, so that finally, from want of confirmation, these fell out of favour, and with them fell the more valuable parts of the Pythagorean teaching.

The general faith of the Pythagoreans in the all-importance of numbers in the scheme of nature led them to imagine that the distances of the various planets from the central fire must stand in the ratio of simple numbers, and so must correspond to the more harmonious intervals in the musical scale. Thus they said that 'the whole heaven is harmony and number', and believed that the planets produced music, inaudible to us, as they moved in their orbits—the 'harmony of the spheres'.

ANAXAGORAS. Anaxagoras of Clazomenae (*c.* 500–428 B.C.) was a rich man who neglected his possessions so as to devote himself to astronomy, saying that the object of being born is 'to investigate the sun, moon and sky', and ultimately got into trouble for his rationalist views. He discovered the cause of the phases of the moon, maintaining, so Aëtius tells us, that the moon's obscurations month by month result from its following the sun which illuminates it, while its eclipses occur whenever it falls within the shadow of the earth. Plutarch says that 'Anaxagoras was the first to put in writing, most clearly and most courageously of all men, the explanation of the moon's illumination and darkness'.

Cleomedes, a Greek mathematician who wrote in the second or third century A.D., tells us that the explanation did not

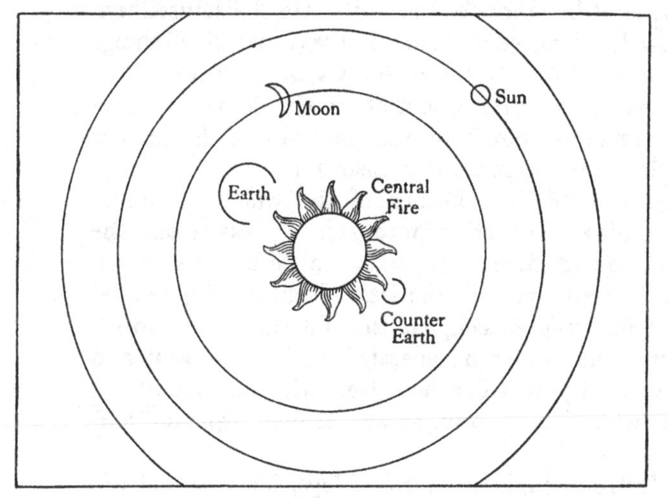

Fig. 13.

escape criticism.* Eclipses were said to have occurred when both the sun and moon were visible above the horizon, and this was thought to disprove the explanation of Anaxagoras.†

* Κυκλικῆς Θεωρίας Μετεώρων.

† But Cleomedes says he doubts the alleged facts, suspecting that they were mere inventions by 'persons who desired to cause perplexity to the astronomers and philosophers'. In any case, he says, if such eclipses did occur, they can be adequately explained by refraction in the earth's atmosphere; this makes it possible to see both the sun and the moon when they are really below the horizon. 'It might possibly happen, in a moist and thoroughly wet condition of the air, that the visual ray should, by being bent, take a direction below the horizon and there catch the sun just after its setting, and so receive the impression of the sun's being above the horizon.'

Anaxagoras expressed other rationalistic and materialistic views, refusing to see anything wonderful or divine in the pageant of the heavens, and maintaining that the heavenly bodies were of the same general nature as the earth, except that they had become incandescent through rotation. He thought that the sun was a vast mass of incandescent metal, larger than the Peloponnese, while the moon had valleys and mountains on it like those of the earth. He conjectured that the universe had 'started as a chaotic mass in which all things were mixed together'. In this a vortex was generated, which spread to ever wider circles, so that air, clouds, water, earth and stones separated out in turn as the result of the circular motion, the heaviest remaining near the centre. Finally, 'in consequence of the violence of the whirling motion, the surrounding fiery ether tore stones away from the earth and kindled them into stars'—a cosmogony which had much in common with the later 'Nebular Hypothesis' of Laplace (p. 239). Anaxagoras thought that other worlds besides our own had been generated in the same way, and were inhabited by men like ourselves, who had cities and cultivated fields like our own, as well as their own suns and moons.

These doctrines explained many things, but they did not prove popular when Anaxagoras expounded them in Athens. Plutarch tells us that the book of Anaxagoras was but little esteemed; it 'circulated in secret, was read by few, and was cautiously received'. We have already noticed how Plato received it. Finally, the Athenians decided to prosecute Anaxagoras for impiety and atheism; he was trying to take away their gods—helpful and friendly beings, on the whole, to whom they could look for help and comfort, and who were susceptible to their entreaties and even to their bribes. Aratus* (p. 13) writes: 'Every way we stand in need of Zeus. We are even his offspring; he, in his kindness to man, points out things of good omen, rouses the people to

* *Phaenomena.*

labour, calling to their minds the needs of daily life, tells them when the soil is best for the labour of the ox and for the pick, and when the seasons are propitious for planting trees and all manner of seeds.'* The average Greek was reluctant to surrender such friendly gods for masses of inanimate earth and metal.

Others, more enlightened, found this array of anthropomorphic gods unsatisfying and, as Xenophanes said, 'by seeking, find in time what is better—one god, the greatest among gods and men, like mortals neither in form nor in thought, but all-seeing, all-hearing and wholly thought. Without toil he sways all things by the thought of his mind and abideth ever in the selfsame place, moving not at all.' But these were equally unwilling to accept a rationalist interpretation of the phenomena of the skies. As Plutarch wrote: 'In those days they refused to tolerate the natural philosophers and stargazers, as they were then called, who presumed to fritter away the deity into unreasoning causes, blind forces and unnecessary properties. Thus Protagoras was exiled and Anaxagoras was imprisoned and was with difficulty saved by Pericles.'

There is some doubt as to what actually happened to Anaxagoras. One account says he was convicted and banished from Athens, only the intervention of Pericles saving him from death, while another says he was acquitted, but nevertheless found it prudent to leave Athens and return to his native Ionia. In either case it is clear that the time for rationalism in human thought had not yet arrived. Instead there started an age-long conflict between religion and science; religion had declared war and initiated that persecution of science which was unhappily to recur so often and figure so largely in the histories of both. In the case of Anaxagoras we see the conflict in its earliest, simplest and crudest form, and its very simplicity and remoteness from present-day conditions make it particularly easy to understand.

* Translation by Sir T. L. Heath, *Greek Astronomy*, p. 112.

PLATO. Astronomy, like physics, was unfavourably influenced at this time by the reactionary attitudes of Plato and Aristotle. Plato's scientific beliefs were not based on observation or knowledge, but simply on his individual views as to what was most appropriate. The universe, he thought, must have been shaped to fit human needs and desires. God must be good, and so must have constructed the most perfect of all possible worlds for us to live in. As the most perfect of all shapes is the sphere, He must have made the universe spherical. So also, as the most perfect curve is the circle, He must have made the planets move in circles. Motion, being of divine origin, must be perfect in its regularity—so that it was a great trouble to Plato that no perfect regularity could be discerned in the planetary motions; he is said to have urged all serious students to try to discover what set of uniform and ordered movements would account for the observed movements of the planets.*

Through most of his life, Plato took it for granted that the earth must stand at the centre of the universe, but he seems to have wavered on this in his later years, when, according to Plutarch, he 'regretted that he had given to the earth the central place in the universe, which was not appropriate to it'. Consideration of the supposed central fire now made him 'regard the earth as placed elsewhere than at the centre, and think that the central and chiefest place belongs to some worthier body'. But he remained unshaken in his belief that the plan of the universe could be better discovered from general principles than from observation, and in his only scientific dialogue—the *Timaeus*, the weakest of them all—he tries to discover the plàn from the wholly gratuitous assumption that the structure is like that of a man—the macrocosm must, he thinks, resemble the microcosm.

Nevertheless we learn, again from Plutarch, that his interest in astronomy redeemed it from the reproach of atheism, and made it a respectable subject for study: 'Through the brilliant

* Simplicius, *De Caelo*.

repute of Plato, the reproach was removed from astronomical studies, and access to them opened up for all. This was because of the respect in which his life was held, and because he made natural laws subordinate to the authority of divine principles.'

ARISTOTLE. The general attitude of Aristotle was similar. His observing powers, which were so successful in biology, failed him as completely in astronomy as in physics. Like Plato, he tried to deduce the plan of the universe from general principles rather than from knowledge, and thought that it must necessarily be modelled on the perfect figures of the sphere and circle. He saw the universe as a system of concentric spheres, all having the earth as their common centre. Outside the sphere of the earth came the sphere of the ocean, beyond this the sphere of the atmosphere, and beyond this the sphere of fire. Thus there were spheres of the four elements in turn—earth, water, air and fire. Beyond the sphere of fire came other spheres carrying the moon, sun and the five known planets, and finally, beyond all, the sphere of the fixed stars. Contrary to the teachings of the atomists, Aristotle thought that some driving power must continually be at work to keep the various spheres and their attached planets in motion, and so postulated another sphere, external to all the others, to provide the needed driving power—the Prime Mover, which Aristotle identified with God himself; it made all the stars and planets move in their various spheres at a uniform speed, 'as a beloved moves a lover'.

But Aristotle was extremely tolerant, and quite realised that other views were tenable. 'If those who study this subject form any opinion contrary to that we have stated we must indeed respect both parties, but be guided by the more accurate.'

In his *Meteorologica* he conjectures that 'the bulk of the earth is infinitesimal in comparison with the whole universe', and goes on to say that 'it is absurd to make the universe in process of change because of so small and trifling changes on earth, when the bulk and size of the earth are surely as nothing in comparison with the whole universe'. He also makes a reasoned defence of the doctrine of the central fire, but it is a typical instance of his reaching wrong conclusions through arguing from faulty and unscientific principles. After quoting the Pythagorean view that 'at the centre there is fire, and the earth is one of the stars' [i.e. planets], he continues: 'Many others might agree that we ought not to assign the central place to the earth, looking for confirmation to theory rather than to observed facts', the theory being that as fire is more honourable than earth, it deserves, and so must have obtained, the more honourable place. 'Arguing from these premisses, they think it is not earth that lies at the centre of the sphere, but rather fire (p. 59).'

EUDOXUS. Leaving this backwater and returning to the main stream of astronomical thought, the first astronomer of note that we meet is another Pythagorean, Eudoxus of Cnidus (408–353 B.C.). He was a good observer, and made very accurate observations on the motions of the planets.

We have seen how Plato had propounded the problem of finding what set of uniform and ordered circular movements would account for the observed planetary motions. Eudoxus's efforts to solve this problem led him to propound a cosmology which was in many ways retrograde. His Pythagorean predecessors had already set the earth moving through space like the other planets; Eudoxus not only put it back at the centre of things, but made it stand still there. Round this fixed centre he supposed that a number of spheres revolved. The outermost of these was simply that sphere of Aristotle's which had the fixed stars attached to it; to the inner spheres no stars or planets were directly attached, but other spheres, to which were attached yet other spheres, and so on. To the final spheres of this series were attached the sun, moon and the five planets, which thus revolved round the central earth in a highly complicated way. To fit his observations, Eudoxus found he needed three spheres each for the sun and moon, and four for each of the planets, making a total of twenty-seven spheres. Then the more exact observations of his pupil Callippus showed that twenty-seven spheres were inadequate; thirty-four were now needed. Here we have the germ of the complicated system of cycles and epicycles which, under the leadership of Ptolemy, was to dominate and harass the astronomy of the next 2000 years.

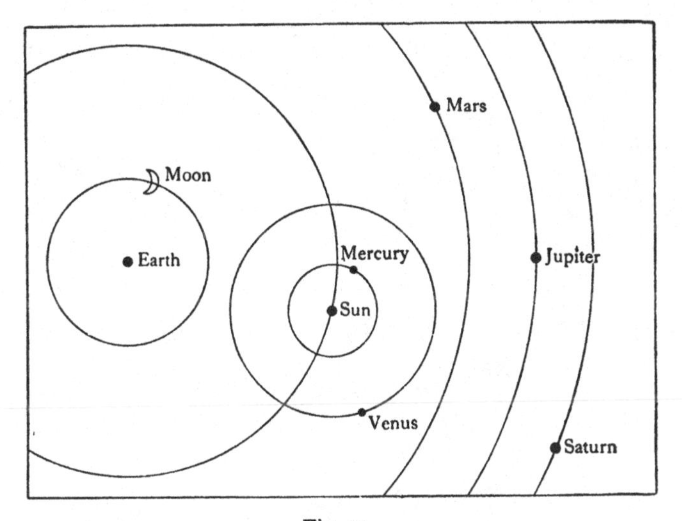

Fig. 14.

During all this time, explorers continued to explore the surface of the earth and notice how the length of day varied from place to place, being dependent on the latitude but not on the longitude. This was supposed to indicate that the earth was spherical in shape. Finally, Ecphantus, one of the last of the Pythagoreans, asserted that this sphere turned on its own axis.

About 350 B.C., Heraclides of Pontus (388–315 B.C.) taught similar doctrines, and added that while the sun and major planets revolved round a fixed earth, Venus and

Mercury revolved round the moving sun (Fig. 14)—an anticipation of the scheme Tycho Brahe was to put forward 1900 years later.

Thanks to Ecphantus and Heraclides, astronomy had now acquired the idea of an earth which was so little fixed that it could rotate under a canopy of fixed stars, and of planets that could revolve round the sun.

CHAPTER III

SCIENCE IN ALEXANDRIA

(332 B.C.–A.D. 642)

THE three centuries we have just had under discussion formed a sort of intellectual 'golden age' in which science made more progress than in three millennia of Babylon and Egypt. But as this period approached its end, a change set in, and by the middle of the fourth century B.C., Greek culture had definitely begun to decline, and Greek science with it. A few years later, the decline was accelerated by the invasion and military conquest of the country by Alexander the Great. Yet events which seemed to be disastrous to science at the time may perhaps have been a piece of good fortune in disguise.

For Alexander now decided to celebrate his victories and consolidate his empire by building a new capital which was to be the most magnificent city in the world. He chose a site on the flat lands where the Nile ran into the sea, and called the still unborn city Alexandria, after himself.

He died in 323 B.C., his grandiose scheme still incomplete, and his kingdom was divided among all who could lay hands on a piece of it. Egypt fell to the lot of one of his generals, Ptolemy, who chose the still unfinished Alexandria as his capital and, more ambitious even than Alexander, aspired to make it the world's capital not only for government and commerce but for culture and intellect as well. To this end he chose a site adjoining his palace, and on it began to build a 'Museum' or Temple of the Muses, which was roughly the equivalent of a modern university. Such was the origin of the city which was to replace Athens as the cultural capital of the Mediterranean world, and of the university which was to provide a home for science for a thousand years to come. These thousand years form the subject of the present chapter.

By about 300 B.C., the university was ready for occupation, and Ptolemy proceeded to staff it with the most eminent scholars of the time; many came to it from Athens, and by so doing carried the torch of learning a step back from west to east. When Ptolemy died in 283 B.C., his successor Ptolemy II, no less ardent to make Alexandria the cultural centre of the world, established the famous library which was accounted one of the seven wonders of the world. It was divided into the four departments of literature, mathematics, astronomy and medicine, each with its own librarian or curator, and is said to have accumulated no fewer than 400,000 manuscripts in the first forty years of its existence.

Fortune alternately smiled and frowned on science in its new home. First there was a series of brilliant successes, made possible in part by the official support of the reigning dynasty, but in part also by a change of method which accompanied the move from Greece to Egypt—a change, as we shall see, from dreamy speculation about the universe in general to precise attacks on clear-cut problems.

Then came a time when science began to wilt again—the period of stagnation before the Christian era. The spirit of progress seemed to have deserted science, in part because many subjects of investigation seemed to have reached their natural endings, and nothing new was found to take their place; discoveries now gave place to comments, criticisms and reviews of past triumphs.

External influences also became less favourable. After governing Egypt for nearly 300 years, the Ptolemaic dynasty came to an end in 30 B.C. with the death of Cleopatra, when the Romans defeated the native Egyptian troops and took over the administration of Egypt. The Romans were great soldiers, great administrators and lawgivers, great engineers and mechanics in an unimaginative practical way, but they were barely even sympathetic to science; their world was the world of affairs, and not of abstract thought. Thus the coming of Alexandria under Roman rule might well have proved disastrous to science, yet the result was not bad. Showing their usual tolerance towards their subject races, the Romans allowed the Greek language and a general Greek atmosphere to prevail in Alexandria, so that life soon resumed its normal course and the University, again full of students, once again became a centre of learning and research.

The real danger came later and from quite another quarter. Christianity, after starting from the humblest of beginnings, conquered the Mediterranean world more thoroughly than ever the Roman legions had done. The Roman conquerors had introduced a new technique of government, but the Christian conquerors brought with them a new technique of life and a revolutionary conception of human aims and destiny—how revolutionary it is hard for us of to-day to understand. Their citizenship was in heaven, their life here only a preparation for a future life elsewhere, so that they saw the world of matter only as a prison-house, and the vault of heaven only as a veil; both were transitory and utterly insignificant in comparison with what lay beyond. Within the lifetime of some of them, a day was to come when the stars would fall from heaven, and the sky be rolled back like a scroll to reveal a Judge seated on his Throne. 'Then God whom Jesus had declared to be the loving Father would change his character, reverting to the ferocity and tyranny of his Old Testament habits: even Jesus himself who had once prayed 'Father, forgive them' would now lay aside mercy and deal out justice and vengeance: sinners for whom he had formerly sought as a shepherd for his lost sheep would now be flung into hell and there would suffer endless flames and torments—a spectacle to increase the beatitude of heaven.'* Tertullian had written: 'How shall I admire, how laugh, how rejoice, how exult, when I behold...so many sage philosophers blushing in red-hot flames with their deluded scholars.' What would it profit a man on that last day of wrath that he had spent laborious years in examining how the

* C. E. Raven, *Science, Religion and the Future*, p. 21.

bars of the prison-house were made, or in studying the heavenly veil that had already passed away? Surely it was better to prepare for the judgement to come?

Holding such beliefs, the Christians could hardly be sympathetic to the study of science, especially as many of them were narrow fanatics; their religion was their all and, unlike the paganism it was supplanting, it knew nothing of tolerance or of magnanimity towards those of other opinions. This mattered little at first, for the Christians were few and uninfluential. Even at the beginning of the fourth century, only a small fraction of the population was Christian;* the pagan writers barely mentioned their existence, even the great moralists such as Seneca and Marcus Aurelius either passing them over in silence or speaking of them with contempt.

Then came the year 312, a landmark in human history, when Constantine the Great, the illegitimate son of a Roman officer and a Serbian innkeeper, who had been elected Emperor of Rome by the army in the field, suddenly embraced the Christian religion.† In 390 the pagan religion was forbidden by edict throughout the Empire, and henceforth Christianity reigned supreme, save in out-of-the-way country places, where the simple villagers would still assemble to sing hymns and offer modest sacrifices to the gods of their forefathers.

Twenty years later Rome was captured by Alaric and his barbarians, and when these too embraced the Christian faith, the 'dark ages' fell upon Europe—the ages of domination of all human thought and of most human activity by the priesthood, ages which 'should probably be placed, in all intellectual virtues, lower than any other period in the history of

* According to Bury, only about a fifth. *History of the Later Roman Empire*, I, 366.

† He does not appear to have done this from any moral conversion or mental conviction, but merely because his use of a Christian emblem seemed to bring him victory in the field. Eusebius relates that he and his troops saw a flaming cross in the sky with the inscription ἐν τούτῳ νίκα (in this conquer), and that, having put this emblem on his banners, he won four victories in rapid succession.

mankind. A boundless intolerance of all divergence of opinion was united with an equally boundless toleration of all falsehood and deliberate fraud that could favour received opinions. Credulity being taught as a virtue, and all conclusions being dictated by authority, a deadly torpor sank upon the human mind, which for many centuries almost suspended its action'.*

In the present chapter our task will be to trace out the fortunes of science from the time of its decline in Greece and rise in Alexandria to the time when this deadly torpor gripped the human mind, a period of nearly a thousand years.

The scene will be laid almost exclusively in Alexandria, for in spite of the various disadvantageous influences at work, Alexandria had established itself so firmly as the cultural centre of the world that nearly all the great scientists of the next millennium either taught or studied there, or both. The scientific spirit exercised itself mainly in the two subjects of mathematics and astronomy. The Alexandrian mathematicians included some of the greatest that the world has seen—Euclid, Archimedes and Apollonius. The same is equally true for astronomy, the great names here being Aristarchus, Eratosthenes, Hipparchus and Ptolemy. Let us now examine Alexandrian science in detail, commencing with mathematics.

MATHEMATICS IN ALEXANDRIA

EUCLID. The first of the great Alexandrian mathematicians, Euclid, was born about 330 B.C., probably of Greek parents,† and died about 275 B.C. We do not know where he was educated, but some think that Athens is indicated, both by his writings and by his knowledge of the works of Plato. He became curator and librarian of the mathematical department of the Alexandrian Library, and he taught as well.

By far the most famous of his works is his *Elements of Geometry*, which determined the way geometry was taught in

* Lecky, *History of European Morals*, II, 13.

† But see p. 25.

our schools until quite recently. We do not know what purpose the book was designed to serve—a text-book for students, or a compendium of geometrical knowledge, or a scholar's effort to demonstrate that the facts of geometry are inevitable truths which can be deduced from axioms of indisputable validity. Actually it serves all three purposes very well, as perhaps it was meant to, but it is the last which interests us most to-day.

For the modern geometer does not think of the axioms as being indisputably true. He agrees that if they are true, the propositions follow as a matter of pure logic. But he regards the axioms, and especially the famous twelfth axiom,* as specifying properties of space. He has to deal with many kinds of space, but there is only one—Euclidean space, he calls it—in which the twelfth axiom is universally true. In this space, but in no other, the theorems of Euclid, as, for instance, the famous theorem of Pythagoras, are invariably true. The properties of other kinds of space are most easily specified by stating the manner and extent in which the theorem of Pythagoras fails in them. All this has been brought into the field of practical science of late, because the theory of relativity (p. 294) depicts the world as existing in a space in which Euclid's axioms are not generally true.

The *Elements* consists of a coherent treatise of twelve books, in which a sequence of propositions is deduced by strict logic from the axioms just mentioned, together with a thirteenth book of disconnected oddments forming an appendix. Possibly, as de Morgan once suggested, the whole work was a product of Euclid's old age, which death prevented his ever putting into final shape. It is largely a compilation. Many of its propositions appear in an earlier *History of Mathematics* which Eudemus wrote when Euclid was only about ten years old, while some of its contents were certainly known to the

* 'If a straight line meets two straight lines so as to make the interior angles on one side of it together less than two right angles, then these straight lines will meet if continually produced on the side on which are the angles which are together less than two right angles.'

Pythagoreans—as, for instance, the theorem that $\sqrt{2}$ is incommensurable, which Euclid gives twice (*Elements*, book x, props. 9 and 117). Many of Euclid's proofs are tedious, slow-moving and obvious, but others show great ingenuity. The following is an example.

We know that numbers can be divided into the two classes of composite and prime, a composite number being one which is the product of smaller factors, as, for instance, 6 (which equals 2×3) and 8 (which equals $2 \times 2 \times 2$), while a prime number is one which cannot be so split up, as for instance 5 or 7. If we examine the first six numbers after 1, we find that two-thirds of them are prime, namely, 2, 3, 5 and 7. If we examine twelve numbers instead of six, the proportion of prime numbers falls to one-half, the primes being 2, 3, 5, 7, 11 and 13. If we take 24 numbers, it further falls to three-eighths; with 48 numbers it is reduced to five-sixteenths; with 96 numbers to one-quarter, and so on. The further we go, the smaller the proportion becomes, the reason being that fresh divisors are continually becoming available. The question now arises: If we go far enough, shall we ever come to a range in which none of the numbers are primes? Or, in other words, is there a largest prime, beyond which no other prime numbers exist? It appears to be a terribly intricate problem; if the reader does not think so, let him try to solve it before reading further. Yet Euclid solves it by the simple remark that if there could be a largest prime number N, then the number $(1 \times 2 \times 3 \times 4 \times 5 \times \ldots \times N) + 1$ would have to be both prime and not-prime, which is absurd. It would have to be not-prime because it is greater than N, which we are supposing to be the largest prime number. But it would also have to be prime; no prime number can be a factor of it, since division by any one of the primes 1, 2, 3, ..., N always leaves 1 over as a remainder. Thus the supposition that there is a largest prime N leads to contradictory conclusions, whence it follows that there cannot be a largest prime.

Besides his *Elements* Euclid wrote four other books on

geometry, and also books on astronomy, music and optics, only the last of which has survived. This states the laws of reflection of light accurately; the laws of refraction were not yet known. But Euclid took a wrong view as to the nature of light. The Pythagoreans had taught that light travelled from a luminous object to the eye in the form of particles—an anticipation of the corpuscular theory of Newton and of the present-day particle-picture of light (p. 331). Empedocles had taught that light was a sort of disturbance which travelled through a medium, taking time on its journey—an anticipation of the undulatory theories of the eighteenth and nineteenth centuries (p. 254), and of the wave-picture of to-day. Plato and others had imagined, quite erroneously, that light consisted of rays which travel in straight lines from the eye until they strike an object, which the eye then sees. When we look for an object, they thought, we poke about for it with these rays, just as we might grope about for something in the dark with our hands. Euclid accepts this last alternative, arguing that light cannot proceed from the object into the eye, since if it did, 'we should not, as we often do, fail to perceive a needle on the floor'.

ARCHIMEDES. Greatest of all the Alexandrian mathematicians, and best known after Euclid, was Archimedes (287–212 B.C.). After studying in Alexandria, he returned to his native Sicily where he was finally killed by the Romans in Syracuse, when they took the city after a three years' siege. Like Pythagoras and Plato, he held that learning should be acquired for its own sake and not for gain, or for its practical applications, but as his life fell in times of war, his great mechanical ingenuity had to be turned mainly to military ends. He is said to have set fire, by mirrors and burning-glasses, to the ships which were besieging Syracuse—a story which many doubt*—and to have devised catapults which kept the besiegers away from the walls of the city. Among his more

* On this see W. W. Rouse Ball, *A Short Account of the History of Mathematics*, p. 67.

peaceful inventions were the 'screw of Archimedes', a device for raising water which was in use in Egypt until fairly recent times, and a cog-wheel and screw arrangement for launching ships. But he is best known for his method of measuring the specific gravities of substances. He put a known weight of the substance into a vessel which was already full of water, and weighed the water which flowed over the rim. If, for instance, he put 12 lb. into the vessel, and found that 1 lb. of water overflowed, he knew that his mass of substance weighed 12 times as much as an equal volume of water, so that its specific gravity was 12. A well-known story records how he detected in this way the fraud of a goldsmith who had adulterated gold which had been given him to make a crown. It adds that he thought of the method while in the baths and ran through the streets crying εὕρηκα, εὕρηκα.

His work in mathematics was of immense range and variety. Many of the common formulae of geometry are attributed to him—πr^2 for the area of a circle (where π is the ratio of the circumference to the diameter), $4\pi r^2$ and $\frac{4}{3}\pi r^3$ for the surface and volume of a sphere, and the corresponding formulae for cones and pyramids.*

Archimedes also arrived at a very good approximation for the value of π, using what was known as the 'method of exhaustions'. The smallest square which can completely enclose a circle of radius r is of area $4r^2$, while the largest square which can be enclosed in the circle is of area $2r^2$ (fig. 15). Plainly, then, the area of the circle must be something between $2r^2$ and $4r^2$. If we had drawn regular hexagons instead of

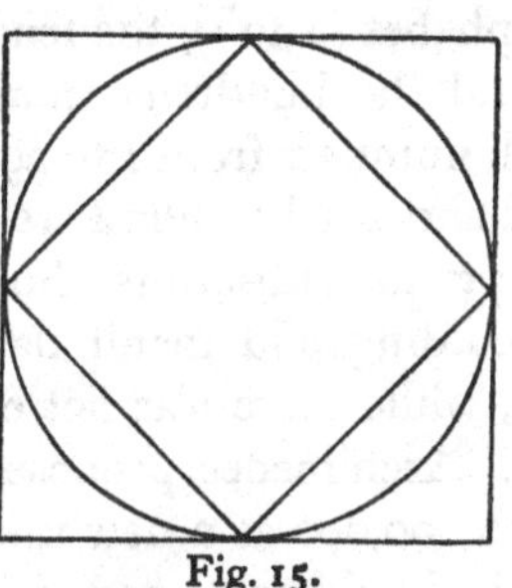

Fig. 15.

* Archimedes says that the formulae for the volumes of a pyramid and a cone were first given by Democritus, but without proof, and that proofs were first given by Eudoxus. But the formula for the volume of a pyramid is to be found in the Moscow Papyrus (p. 12) of at least a thousand years earlier.

squares, we should have found the closer limits $2{\cdot}598r^2$ and $3{\cdot}464r^2$, while octagons would have given the still closer limits $2{\cdot}828r^2$ and $3{\cdot}314r^2$. The more sides the polygons have, the more closely they will grip the circle, and the narrower the limits they will provide. With 96-sided polygons, the limits are found to be $3{\cdot}1395r^2$ and $3{\cdot}1426r^2$, whence of course the value of π must lie between 3·1395 and 3·1426. Archimedes used a polygon of 96 sides, but introduced certain numerical approximations* which brought him to the result that π must lie between $3\frac{10}{71}$ (or 3·1408) and $3\frac{10}{70}$ (or 3·1429). The true value of π is of course 3·1416.

Archimedes also wrote a number of small treatises on various subjects, such as the principles of the lever and pulley, on spirals (especially the well-known 'spiral of Archimedes'), on the area of the parabola, on arithmetic, and so forth. Most of these are lost, but the two following examples of his arithmetic which have survived are of interest as showing the high level to which he attained.

The Greeks were still using letters to denote numbers, and a variety of systems was in use. In Alexandria they represented the numbers from 1 to 9 by the first nine letters of the Greek alphabet (α to ι), the tens from 10 to 90 by nine more letters, and the hundreds from 100 to 900 by yet another nine.† All numbers from 1 to 999 could be represented in this notation, and further numbers up to 99,999,999 by adding superscripts and subscripts. But the clumsiness of the system made recording and manipulation difficult even for small numbers, while there was not even a notation for very large numbers. Archimedes proposed to deal with the latter by taking 100,000,000 as a new unit, and that the square, cube, etc., of this should be treated as additional units 'of the second, third and so on' orders. If, as in modern mathematics, we denote a 1 followed by any number n of zeros by 10^n, then

* Sir T. L. Heath, *Manual of Greek Mathematics*, pp. 295–309.

† As the Greek alphabet only contained 24 letters, it was necessary to supplement these by two obsolete Greek letters, and one Phoenician letter.

Archimedes proposed taking 10^8 as his first new unit, while the others would be 10^{16}, 10^{24}, 10^{32}, and so on—just as we take a million as a sort of unit, and also speak of billions, trillions, quadrillions, and so on. To illustrate the working of his proposed system, he calculated the number of grains of sand that would be needed to fill the universe. Assuming that 10,000 grains of sand can be put into a sphere one-eightieth of a finger-breadth in radius, and that the diameter of the universe is less than ten thousand million stadia (about a thousand million miles, which is only slightly more than the diameter of Jupiter's orbit), he calculates that the number will be less than 10^{63}.* Here we have the prototype of the kind of calculations which figure so largely in modern astronomy.

Archimedes points out that the different units 10^8, 10^{16}, 10^{24}, etc., form what we now describe as a geometrical progression, and makes the pregnant remark that the product of the mth and nth units is equal to the $(m+n)$th unit, or, in modern language, that $x^m \times x^n = x^{m+n}$. Here we have the first known statement of the law of indices, the germ out of which calculation by logarithms was to emerge 2000 years later.

The second example is of a very different type. Archimedes proposed the following problem as a challenge to the mathematicians of Alexandria. 'The sun had a herd of cattle of different colours—piebald, white, grey and dun. The number of piebald bulls was less than the number of white bulls by $(\frac{1}{2}+\frac{1}{3})$ times the number of grey bulls, and less than the number of grey bulls by $(\frac{1}{4}+\frac{1}{5})$ times the number of dun bulls, and less than the number of dun bulls by $(\frac{1}{6}+\frac{1}{7})$ times the number of white bulls. Furthermore, the number of white cows was $(\frac{1}{3}+\frac{1}{4})$ times the total number of grey cattle (bulls and cows together), while the number of grey cows was $(\frac{1}{4}+\frac{1}{5})$ times the number of dun cattle, the number of dun cows was $(\frac{1}{5}+\frac{1}{6})$ times the number of piebald cattle, and the number of piebald cows was $(\frac{1}{6}+\frac{1}{7})$ times the number of white cattle. How many bulls and cows were there of various colours?'

* Ψαμμίτης (the Sand-reckoner).

We may think it easy to rewrite these data in the form of simultaneous linear equations which, though complicated, will not be difficult to solve; but such methods were unfamiliar to Archimedes, and, in any case, the arithmetic is far from simple. There is, of course, no unique solution, since the data only determine the proportions, and not the absolute size, of the herd. Archimedes gave the solution:

piebald	331,950,960	bulls,	and	435,137,040	cows,
white	829,318,560	bulls,	and	576,528,800	cows,
grey	596,841,120	bulls,	and	389,459,680	cows,
dun	448,644,800	bulls,	and	281,265,600	cows,

in which all the numbers are multiples of 80, so that a simpler solution is obtained by dividing them all by 80. It is almost unthinkable that Archimedes could have manipulated numbers of this magnitude in terms of the clumsy system of numeration then in common use, so that he may have reached his results by some other and private system, and then translated it back to the common system to announce it to the world. The example we gave before this suggests that he may even have had a system not unlike our own of to-day.

Archimedes was undoubtedly the greatest of all the Greek mathematicians, and would have appeared still greater had not the accidents of war and siege restricted his activities and shortened his life. When the Romans finally took Syracuse, the soldiery were ordered to spare his life and home, but, either through accident or design, this was not done. The Roman conquerors built him a splendid tomb, on which was engraved a diagram of a cylinder circumscribing a sphere, to commemorate the way in which he calculated the area of a spherical surface. It had been his own wish to be buried under such a tomb.

HERO OF ALEXANDRIA. From Archimedes, our thoughts turn naturally to Hero, another Alexandrian mathematician. His date is uncertain, but he was probably a full century, and possibly several centuries, after Archimedes.* He possessed much the same kind of mechanical ingenuity as Archimedes, although to a lesser degree, and displayed the same neat mathematical craftsmanship. But Archimedes was a mathematician by choice, and a mechanic and inventor only by necessity, whereas in Hero the parts seem to have been reversed. He invented a great number of conjuring tricks and mechanical toys, one of the more noteworthy being a steam engine. Steam was produced by boiling water, and passed into a hollow tube which could rotate about an axis. Four nozzles led out of this tube to the outer air, all so bent that the steam escaping through them set the tube rotating by its back-pressure—in the manner of the jet-propelled aeroplane. Here is the first known instance of steam pressure being used to translate chemical energy of burning fuel into energy of motion, the principle underlying the steam engine of to-day. Hero is also said to have devised the first penny-in-the-slot machine recorded in history.

On the abstract side, Hero did some good mathematical work, his studies in optics being of special interest. Euclid had stated that when light is reflected from a smooth surface, the angle of incidence is equal to the angle of reflection. Hero showed that the same law can be put in the alternative form that the light follows the shortest path from point to point, subject to the condition that it must strike the mirror at some point of its journey. Thus if *ABC* in fig. 16 is the actual path, this is shorter than *AB′C* or *AB″C*, or any other similar path. Hero does not appear to have attached any special

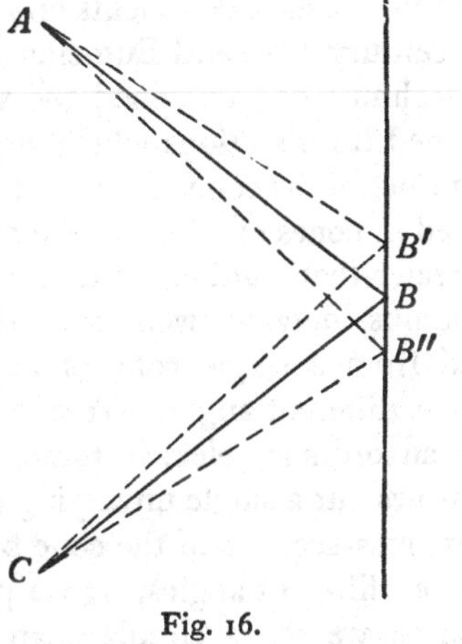

Fig. 16.

* Sir T. L. Heath, *Manual of Greek Mathematics*, p. 415.

importance to this result, and little suspected that he was in fact introducing a new and far-reaching principle which was to develop into one of the most important of all the methods of mathematics.

APOLLONIUS. We have seen how Menaechmus introduced the conic sections into mathematics, but made no great use of them. Euclid and Archimedes also worked on the curves, but most of their writings are lost. Then came Apollonius (260–200 B.C.), a mathematician who had studied at Alexandria for many years and probably taught there as well, to infuse new life into the study. In brief, he did for the conic sections what Euclid had done for the circle a hundred years earlier, writing a treatise which was so comprehensive that centuries were to pass before any substantial additions were made to the subject. It contained about 400 propositions, and was divided into eight books: we know their contents fairly well, for seven of the eight are still extant, four in the original Greek text, and three more in Arabic translation. Besides this, we have comments on the whole work by Pappus (fourth century A.D.) and Eutocius (sixth century A.D.).

Menaechmus had imagined the various conic sections to be obtained from sections which were always at right angles to the surface of the cone, and found that three different curves originated in cones of which the angles were respectively less than, greater than, and equal to, a right angle.

Apollonius now showed that all three curves can be obtained from a single cone of any angle, by cutting the sections at different angles. We can see this for ourselves by flashing an ordinary electric torch on a floor or wall. The torch throws out a single unvarying cone of light, and we see different cross-sections of the cone by letting the light fall on the floor at different angles. If we point the torch vertically downwards, we see a circular patch of light on the floor showing that the curve of cross-section is a circle. But if we turn the torch through a small angle, the patch of light becomes elongated, and the curve of section is an ellipse (or elongated circle). If we turn the torch until it points horizontally, the curve of section becomes a parabola. If we turn the torch still farther in the same direction, so that it now points slightly upwards, the curve of section becomes a hyperbola. It is usual to think of the cone as extending in both directions from its vertex, a state of affairs we cannot reproduce with our torch. When we think of the cone in this way, the hyperbola consists of two detached curves as in fig. 12 (p. 40).

Apollonius also gave their present names to the conic sections, parabola meaning 'the application', ellipse 'the deficiency', and hyperbola 'the excess'.*

The conic sections have now acquired a special importance from their frequent occurrence in nature, but the Greeks knew nothing of this, imagining that most natural motions must necessarily take place in circles. This view had to be abandoned when Kepler found in 1609 that the planets moved in conic sections, and when Newton showed in 1687 that they had to do this if their motion was determined by the gravitational attraction of the sun. Conic sections became even more important when modern physicists began to picture the atom as electrified particles moving round attracting centres in conic sections.

While these curves have become important, the Greek method of studying them has fallen into disuse. The method was to construct a chain of propositions, each being deduced by strict logic from those which preceded it—as Euclid did in his *Elements*—and so forming a compendium of results. But such a compendium has now become about as useless as the compendium of arithmetical results which Ahmes transcribed in his papyrus has become, now that modern methods

* The appropriateness of the names will be seen if the equations of the curves are written in the forms:

$y^2=\alpha x$ (the parabola),

$y^2=\alpha x-\beta x^2$ (the ellipse, y^2 being in deficiency by βx^2),

$y^2=\alpha x+\beta x^2$ (the hyperbola, y^2 being in excess by βx^2).

of handling numbers provide an almost effortless short cut to any result we may want at any moment. The purely geometrical methods of Menaechmus, Euclid and Apollonius have been similarly superseded by what is known as 'analytical geometry'. This is usually said to have been the invention of Descartes (1596–1650) and Fermat (1601–65), and will be explained when we come to their period, but it was probably in use long before their time, and quite possibly even by Apollonius. Its methods are incomparably more direct, more powerful and more certain than the groping methods of Greek geometry.

These latter methods had reached their natural limit with Apollonius, so that geometry lay almost stagnant through the great scientific calm that ushered in the Christian era, and for many centuries after. In the second half of the fourth century A.D., one geometer of outstanding ability appeared in Pappus of Alexandria, but he had the misfortune to be born out of his time, when the interest in geometry was already dead. His only surviving work, Συναγωγή (the Collection), is a compendium of mathematical knowledge, and is of interest because it describes the contents of other books that have since been lost. Mathematicians still associate the name of Pappus with a problem which he propounded in this book, but only partially solved, namely, to find the path of a point which moves so that the product of its distances from a number of lines stays always proportional to the product of its distances from a number of other lines. Euclid and Apollonius had solved certain simple cases of this problem, and Descartes solved it in its general form; indeed, it is said to have been this that led him to the invention of analytical geometry.

DIOPHANTUS. At about the same period as Pappus, we meet another great Alexandrian mathematician in Diophantus, who is usually credited with the introduction of algebraic methods into mathematics, and was certainly the earliest of known writers to make a systematic use of symbols. He employed them to denote powers, equality, the negative sign, and so forth, although it may be that others whose books are now lost may have done so before him.

We have already noticed how Euclid proved theorems of geometry which were equivalent to algebraic formulae, the reason for their geometrical setting being that the Greeks of his time usually thought of quantities in terms of lengths and areas. To take a well-known example, Euclid presented the theorem that

$$(a+b)^2=a^2+2ab+b^2$$

in the geometrical form that 'the square on *AB* (fig. 17) is equal to the sum of the squares on *AC* and *CB*, together with twice the rectangle of sides *AC*, *CB*', and up to the time of Diophantus, it was unusual to announce theorems which did not admit of a geometrical interpretation.* The science of quantities was cramped in geometrical fetters, until Diophantus came to break the fetters, and set it free.

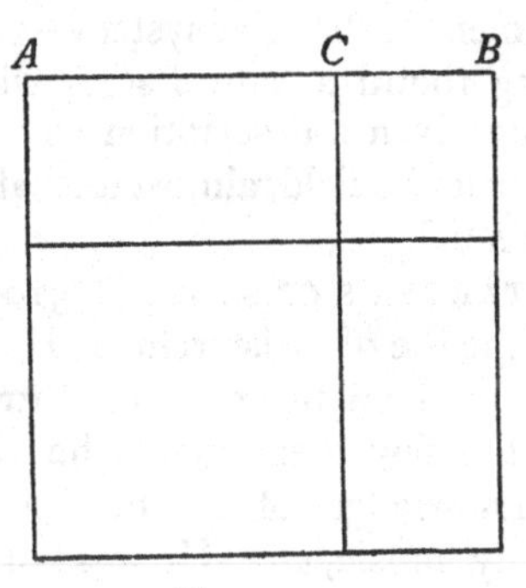

Fig. 17.

Diophantus used his new algebraic methods to solve equations of the first and second degrees, i.e. linear and quadratic equations of the forms

$$ax+b=0 \quad \text{and} \quad ax^2+bx+c=0,$$

and it is of interest that he used just the same methods as are

* Hero had announced that the area of a triangle of sides *a*, *b*, *c*, was

$$\tfrac{1}{4}\sqrt{[(a+b+c)(-a+b+c)(a-b+c)(a+b-c)]},$$

a form of statement which the early Greeks would have found meaningless, since it needed a four-dimensional space for its geometrical representation. But this form of statement was so unusual that Hero apologised for multiplying four factors, where only three could be represented in a diagram.

taught in our schools to-day. He also solved a few simple simultaneous equations, and the very simple cubic equation $x^3+x=4x^2+4$.

ASTRONOMY IN ALEXANDRIA

In the thousand years we now have under review, we find four really great astronomers associated with Alexandria—Aristarchus, Eratosthenes, Hipparchus and Ptolemy. The first is noteworthy as having given the first true description of the arrangement of the solar system—planets, including the earth, revolving round a central sun—while the last is noteworthy as having given a description which was entirely erroneous, and yet held the field, almost unchallenged, until the sixteenth century A.D.

ARISTARCHUS OF SAMOS (*c.* 310–230 B.C.). We know but little of the life of Aristarchus. He was sometimes described as 'the mathematician', but Vitruvius speaks of him as one of the few great men who possessed an equally profound knowledge of all branches of science—geometry, astronomy, music, etc. He was born in Samos and became a disciple of Straton who, as one of the earlier Peripatetics, had probably been in close touch with Aristotle. Straton had tried to explain everything on rationalistic lines, so that it is not surprising that Aristarchus should approach the problems of astronomy from a similar angle.

He was the first to treat astronomical observations in a truly scientific spirit, and to make deductions from them by strict mathematical methods. In a work which is still extant, *On the Sizes and Distances of the Sun and Moon*, he tries to calculate these quantities by pure deduction from observation.

We have seen how Anaxagoras (p. 61) had given the true explanation of the phases of the moon. As the sun and moon move about in the sky, that fraction of the moon's surface which is seen at the earth to be lighted by the sun changes continually. At the moment when precisely half is seen illuminated, the angle *EMS* in fig. 18 must be exactly a right angle. If, then, Aristarchus could measure the angle *MES* at such a moment, he would know the shape of the triangle *MES*, and so could calculate the relative distances of the sun and moon.

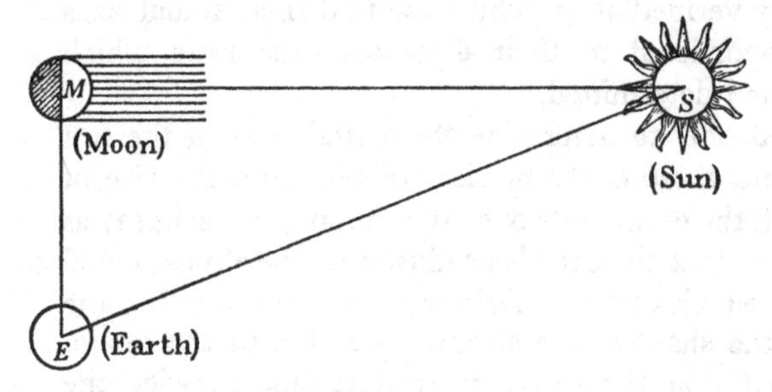

Fig. 18.

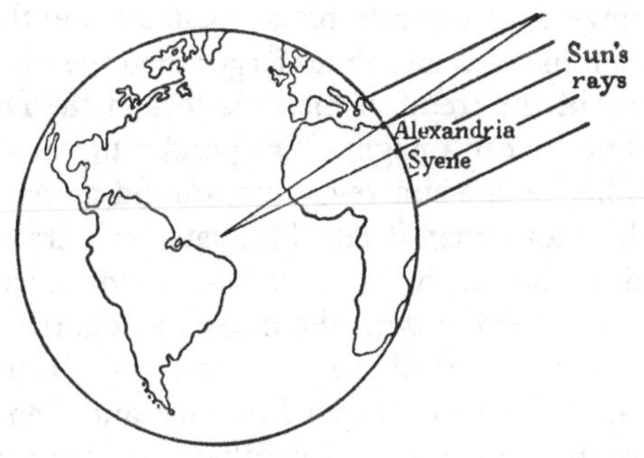

Fig. 19.

Such was his ingenious and perfectly sound method. But the moment of exact half-moon is difficult to estimate, and Aristarchus estimated the angle *MES* to be only 87° at this moment, whereas the true value is 89° 51′. The error was more serious than appears on the surface, because the final result of the calculation turns on the small difference between this angle and 90°. Aristarchus estimated this at 20 times its true value, and concluded that the sun is about 19 times as distant as the moon, whereas the true figure is just 20 times greater. Nevertheless this calculation, inexact though it was, drew attention to the inequality in the distances of the sun and moon. It showed also that the sun and moon must be of very different sizes. They look the same size in the sky, as can be most easily verified at an eclipse, so that their actual sizes must be proportional to their distances—the ratio which Aristarchus had determined.

It remained only to determine the actual sizes of the sun and moon, and these could be determined from the size of shadow which the earth casts over the moon at an eclipse; as the sun is so distant the earth's shadow must be almost equal to the earth which casts it. Aristarchus estimated that the diameter of the shadow was about twice that of the moon, and concluded that the earth must have about twice the diameter of the moon, although the true figure, as we know, is about 4. But, however inaccurate his estimates were, they showed that the sun must be many times larger than the earth.

We know nothing of the trend of his thoughts in the face of this discovery, but we can imagine him pondering on the inherent improbability of the sun revolving around an earth which was so much smaller than itself. Philolaus had already proposed dethroning the earth from its supposed central position, and making it revolve with the other planets round a new centre, the 'central fire' of the universe, while Heraclides had taught that the two planets Mercury and Venus revolved round a centre which was none other than the sun. Why not, Aristarchus may have thought, combine the two suggestions and suppose that all the planets, including the earth, revolve around the sun?

Aristarchus probably saw that if the earth moved in this way, its motion would cause the fixed stars continually to change their directions as seen from the earth, so that the appearance of the sky ought continually to change. Yet no such change was noticed, and he may have seen that this could only mean that the stars are so enormously remote that the earth's motion round the sun makes no appreciable change in their apparent positions. In any case Archimedes wrote a few years later that Aristarchus put forward the hypotheses 'that the fixed stars and the sun remain motionless, that the earth revolves about the sun in the circumference of a circle, the sun lying in the middle of the orbit, and that the sphere of the fixed stars, situated about the same centre as the sun, is so great' that the earth's orbit 'bears the same proportion to the sphere of the fixed stars as the centre of a sphere bears to its surface'.*

By abandoning the usual Greek methods of speculation and reliance on supposed general principles, Aristarchus had attained, almost at one bound, to an accurate understanding of the arrangement of the solar system; he had gained true ideas as to the relatively minute size of the earth, its apparent unimportance as a mere appendage of a far vaster sun, and the insignificance of both in the vastness of space.

In this way astronomy was started on the right road, and we might expect that the rest of the story would be one of rapid progress on scientific lines. Actually it was to be very different. Plutarch tells us that the doctrines of Aristarchus were confidently held, and even violently defended, by Seleucus of Babylon in the second century B.C., but apart from this isolated adherent, we hear of but little support for them until the time of Copernicus and Galileo.

The truth seems to be that such doctrines were simply too far in advance of their time to prove acceptable to either the simple or the learned. The solid, sturdy, unimaginative 'horse-sense' of the average citizen told him it was absurd to imagine that anything so large as the earth could be only a minute fragment of the universe, and still more absurd to imagine that anything so big and solid could be in motion—and if it were, pray, what could produce the immense forces that, according to the mechanical ideas of the time, would be needed to keep it in motion?

* Archimedes, Ψαμμίτης (the Sand-reckoner).

Moreover, we can imagine the average citizen feeling reluctant to surrender his comfortable feeling of consequence as an inhabitant of the most important part of the universe, or as a near neighbour of the gods. And so Cleanthes proposed that Aristarchus should be charged with impiety, as Anaxagoras had been two centuries earlier, and again religious intolerance helped to divert thought from the truth. Astronomy was brought back to essentially the point at which it had been left by Eudoxus, and ideas generally similar to those of Eudoxus were to mould astronomy for 2000 years to come.

ERATOSTHENES (*c.* 276–195 B.C.) was the chief curator of the library at Alexandria and had the reputation of being the most learned man of antiquity. He was nicknamed 'Pentathlos', a tribute to his versatility. He wrote on many subjects, but is best known for his measurement of the dimensions of the earth. The principle was extremely simple, and was not new.

He believed that at noon on midsummer day, the sun was exactly overhead at Syene (the modern Aswan), so that the bottom of a well was directly illuminated by the rays of the sun, and he found by measurement that at the same moment at Alexandria the sun was a fiftieth of a complete circle (or 7° 12′) below the zenith. He believed Aswan to be due south of Alexandria, and concluded that the earth's surface at Aswan made an angle equal to a fiftieth of a complete circle with the earth's surface at Alexandria, whence the circumference of the earth must be 50 times the distance from Syene to Alexandria. Estimating this latter distance to be 5000 stadia, Eratosthenes concluded that the circumference of the earth was 250,000 stadia. Archimedes tells us that it had been previously estimated at 300,000 stadia.

Eratosthenes seems to have subsequently amended his estimate to 252,000 stadia. We do not know what the precise length of the Egyptian stadium was, but if we assume the likely length of 517 ft., the circumference comes out at about

24,650 miles, as against a true value of 24,875 miles. But Eratosthenes seems to have made his measurements only in very round numbers indeed, so that to some extent the accuracy of his final result must be attributed to pure good fortune.*

Eratosthenes is also said to have measured the 'obliquity of the ecliptic', i.e. the tilting of the earth's axis of rotation which causes the seasons, and obtained the value of $\frac{11}{166}$ths of a complete circle, or 23° 51′, whereas the true value at the time was about 23° 46′.

HIPPARCHUS. The next great figure we meet is Hipparchus of Nicaea (*c.* 190–120 B.C.). From the time of Aristarchus on, numbers of astronomers had recorded the positions of the brighter stars relative to certain standard points in the sky. Hipparchus built an observatory at Rhodes, and made similar measurements. His reason was that in about 134 B.C. he had found that the bright star Spica had changed its position by about 2° in the preceding 160 years, and this had suggested the need for a new and more exact list of star positions. He accordingly drew up a list of about 1000 stars, this being the number that can be seen with ease in Egypt, and proceeded to measure their positions with all the accuracy he could command.

He next compared this star list with records of the time of Aristarchus, and also with some still earlier Babylonian records. He may have expected to find that here and there an individual star had changed its position in the sky; what he actually found was a systematic series of changes which indicated that the earth's axis had changed its direction in space; it had not always pointed to the same point in the sky. To repeat the simple analogy we have already used (p. 8),

* Actually Aswan is not exactly on the tropic, as Eratosthenes assumed, but about 40 miles north of it, and it is not due south of Alexandria, but about 180 miles to the east; the difference of latitude between the two places is not 7° 12′ but 6° 53′. The first error of 40 miles would alone result in an error of 2000 miles in the circumference of the globe.

the earth did not spin like what the boys describe as a 'sleeping' top, but was wobbling like a 'dying' top. This phenomenon is known as 'the precession of the equinoxes', and its discovery is usually credited to Hipparchus, although a claim is also made for the Babylonian Kidenas with whose work Hipparchus was acquainted. Hipparchus estimated that the earth's axis moved through an angle of about 45″ a year, but the true value is about 50·2″, so that the earth-'top' requires about 25,800 years to complete a wobble and return to its original orientation. It is a great time, but not enormously long in comparison with historical time, so that within human history the earth's axis must have pointed in directions substantially different from that of to-day. We have already seen how this knowledge can be used to date the naming of the constellations, and, in the same way, if we had not known the date of Hipparchus, we could have deduced it from the positions he assigned to the stars.

He also studied the motions of the sun, moon and planets across the sky, and obtained results of great accuracy, giving the length of the lunar month accurately to within a second, and that of the solar year with an error of only 6 minutes. Indeed, he made good measurements of most of the fundamental quantities of astronomy, and in so doing placed quantitative astronomy on a reasonably exact basis. He tried to devise an arrangement of planetary orbits which should account for the observed motions of the planets across the sky. Most of his writings are lost, but it is likely that his scheme was very similar to that which Ptolemy subsequently gave in his *Almagest*, although perhaps less final in form (p. 93).

He is generally credited with the invention of trigonometry, although his writings on the subject are all lost. He is said to have constructed what we now call a table of natural sines,* and is believed to have discovered the theorem (generally

* This in effect gives the length of the chord subtending a given angle at the centre of a circle.

known as Ptolemy's theorem) which we express in the form

$$\sin (A+B)=\sin A \cos B+\cos A \sin B,$$

and virtually contains the whole of elementary trigonometry.

Hipparchus is also said to have known how to 'solve' spherical triangles, i.e. to calculate all the angles and sides of a triangle drawn on a sphere, such as the surface of the earth, when three of the six are given. This would, for instance, enable the astronomer navigator to calculate the distance between two points of which the latitude and longitude were known. Incidentally, the plan of specifying the position of a place on the earth's surface by its latitude and longitude originated with Eratosthenes, but the corresponding plan for the sky with Hipparchus.

Hipparchus died somewhere about 120 B.C., and after him no astronomer of importance appeared for more than two centuries. In astronomy, as in other subjects, the Christian era opened in a period of scientific stagnation.

PTOLEMY. The first astronomer of consequence that we meet on the other side of the gap is Claudius Ptolemy, who is not known to have been in any way related to the royal house of the same name. He taught and made observations in Alexandria from about 127 to 151 A.D., and is believed to have died about 168. His best known work, the *Almagest*,* did for astronomy what Euclid's *Elements* had done for geometry, and remained the standard book on the subject until the seventeenth century. Like the *Elements* it consists of thirteen books, and it contains much mathematics as well as astronomy. Some of it is original, but much is obviously taken from earlier writers, especially Hipparchus.

Book I, which is a treatise on trigonometry, is noteworthy

* The original title was μεγάλη σύνταξις τῆς ἀστρονομίας, but some Greek translator appears to have changed μεγάλη into μεγίστη which some Arab may have then changed into 'al megiste', and hence the usual name.

for a table of natural sines.* The value of π is given as $3\frac{17}{120}$, or $3{\cdot}14167$, which compares with other values as follows:

Limits assigned by Archimedes
$$\begin{cases} 3\frac{10}{70} = 3 + \frac{1}{7} = 3{\cdot}14286 \\ 3\frac{10}{71} = 3 + \frac{1}{7\frac{1}{10}} = 3{\cdot}14084 \end{cases}$$

Ptolemy's value
$$3\frac{17}{120} = 3 + \frac{1}{7\frac{1}{17}} = 3{\cdot}14167$$

A later approximation
$$3\frac{16}{113} = 3 + \frac{1}{7\frac{1}{16}} = 3{\cdot}1415929$$

True value
$$3{\cdot}1415927.$$

Two other books contain the places of 1022 stars, while others deal with the theory of planetary motions. These, which are the most famous part of Ptolemy's works, definitely put the earth back to the centre of the universe. Eudoxus and Callippus had imagined the planets to be attached to a complicated system of moving spheres; Ptolemy replaced these spheres by a system of moving circles, the general arrangement of which is shown in fig. 20.

In this scheme the sun and moon move round the earth in circular orbits, but the motions of the other planets are more complicated. Out beyond the orbit of the sun is another circular orbit in which nothing material moves—only a mathematical abstraction known as the 'fictitious Mars'. While this is moving round the circle, the real Mars moves in a smaller circle round the fictitious Mars. The big circle in which the fictitious Mars moves is called the 'deferent of Mars', while the smaller circle in which the real Mars moves is called the 'epicycle of Mars', since it is one circle superposed upon another. At some stages of this motion, Mars will be moving in its epicycle in the same direction in which the fictitious Mars moves in its deferent; the motions in the epicycle and the deferent will then reinforce one another, and Mars will appear to move very rapidly across the sky. But at other

* See the footnote to p. 92.

stages, when the motion in the epicycle is in some other direction, Mars will appear to move less rapidly; sometimes the motion in the epicycle will be in exactly the opposite direction to that in the deferent, and Mars will then appear to move backwards. All this fits in well with the observed motion of Mars; it usually moves across the sky in the same direction as the sun and moon, but at times it appears to hesitate in this motion, and occasionally it moves for a short time in the opposite direction.

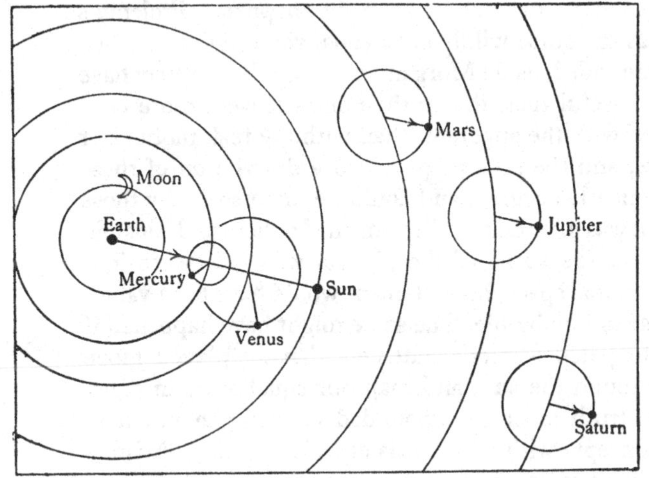

Fig. 20.

Still farther from the earth, Ptolemy proposed similar arrangements of deferents and epicycles for Jupiter and Saturn. There were also deferents and epicycles for Mercury and Venus, but these were made different in character, so as to fit in with the different quality of motion of these two planets. For while Mars, Jupiter and Saturn on the whole drop steadily to the eastward of the sun, Mercury and Venus oscillate round the sun without ever moving far away from it. Ptolemy explained this by supposing that the deferent circles of Mercury and Venus lay between the earth and the sun's orbit, and that the fictitious planets so moved in their deferents as always to lie exactly between the earth and the sun; this made the true planets in their epicycles appear to move round the sun. But it was a highly artificial device, and it seems strange that Ptolemy did not think of making deferents of Mercury and Venus coincide with the orbit of the sun as Heraclides of Pontus had done; even the ancient Egyptians are said to have believed that these planets revolved directly round the sun.

As a representation of what really happens, Ptolemy's scheme was of course wildly erroneous, yet at the time when it was propounded, as de Morgan once remarked, it may have been more useful than the truth. For men were more concerned then with the apparent, than with the real, motions of the planets, and the scheme provided a description of these which was nearly accurate and could be understood by those for whom it was intended. If Ptolemy had anticipated Einstein and said that the paths of the planets were geodesics in a four-dimensional space, his statement would have been valueless because unintelligible. The same might have happened if he had anticipated Kepler's statement that the planets move in ellipses round the sun, and map out equal areas in equal times. The truth must be expounded to each generation in terms of concepts with which it is already familiar. Aristarchus had failed to carry conviction through being too far in advance of his time; Ptolemy, probably with a less penetrating vision than his great predecessor, succeeded because he was nearer to the level of contemporary thought.

Ptolemy also wrote a five-volume treatise on Optics, most of which survives in a twelfth-century translation from Arabic into Latin. In the last of the five volumes he makes a study of the astronomical effects of the refraction of light. He knew that when rays of light passed from one substance to another, as, for instance, from air to water, they are 'refracted' or bent away from a straight course, and he saw that rays of starlight would be bent as they passed from the rare air high up in the atmosphere to the denser air beneath.* This causes a star to appear more directly overhead than it really is, so that for instance, the sun, moon and stars remain visible after they have actually passed below the horizon. Ptolemy describes the results of experiments he had performed on the refraction of light by glass and water, and gives tables of refraction, assuming a law of refraction which is very nearly correct when the angle of refraction is small.†

Ptolemy described two new astronomical instruments, the astrolabe and the mural circle, which were of great use not only at the time but also for many centuries after. He also discussed geography from an astronomical standpoint, explaining the principles of map-making, and agreeing with Hipparchus that observations of latitude and longitude ought to be the first step. But he was not in a position to carry out his own precepts, and could only produce a number of unsatisfactory maps by piecing together odd scraps of information that he had collected from traders and travellers.

A number of other books on optics, astrology, sound and other subjects have been attributed to him, but their authorship is doubtful, and they contain nothing to give their author a fame comparable with that which he acquired from the *Almagest*.

PHYSICS AND CHEMISTRY IN ALEXANDRIA

There was but little to record either in physics or chemistry in Alexandria; the main incident was the rise and fall of the study of alchemy in the third century. The word alchemy has

* This at least is how we express it in modern language. Actually Ptolemy followed Euclid in thinking that rays of light are emanations from the eye, which groped around in space until they fall upon the seen object, so that he would have said that the rays of light are bent as they pass out of the dense lower atmosphere into the rarer air above.

† Ptolemy's law was $\theta' = \mu\theta - \nu\theta^2$, in place of the correct law, $\sin\theta' = \mu \sin\theta$, to which it approximates closely when the angles concerned are small (p. 200 below).

become associated in our minds with all sorts of silliness and fraud, but strictly speaking it ought merely to denote the early form of chemistry. In Alexandria, its practice was a monopoly of the priestly caste, and its secrets were very carefully guarded. Many of them, however, are recorded in a collection of third-century papyri in the Library at Leyden. The general aim of alchemy was the transmutation of base metals into the 'noble' metals, gold and silver. The special aim of the alchemy of third-century Alexandria seems to have been the production of cheap imitations of articles of gold and silver. For instance, a mass of alloy would be made by combining plenty of base metal with a little gold; it would then be shaped and immersed in a mordant salt such as is now used for etching. This would attack the base metal but not the gold, and so would leave a piece of metal which would not only look like gold, but would actually be gold—at any rate so long as no one probed below the surface. There was no real fraud here; the process was almost the exact counterpart of our own electro-plating. Alchemy was practised in Alexandria until about the end of the third century after Christ, when the Emperor Diocletian made it illegal, and ordered all books on the subject to be burned. At first it was of the innocent kind already described, but later there seems to have been some pretence that the base metals were really transmuted into gold by processes of this kind.*

THE END OF THE ALEXANDRIAN SCHOOL

As the fourth century was nearing its end, we come upon the astronomer-mathematician Theon, who wrote a commentary on the *Almagest* and issued a new edition of Euclid's *Elements*, and also his more distinguished daughter Hypatia, the only known woman scientist of antiquity; she wrote commentaries on the conics of Apollonius and on the algebra of Diophantus.

* Dampier, *A History of Science*, pp. 55 ff.

The inspiration of original scientific work had long ago left Alexandria; any original thoughts that the school now produced were philosophical speculations of the mystical, dreamy type,* and its main work was editing, commenting and rehearsing the glories of a bygone age.

The opposition of the Christians to all non-Christian learning was now becoming formidable, but science was too moribund to attract much of it. The Christians were not concerned with science; their all-absorbing interest was in theological controversy. Maintaining that the holding of incorrect theological opinions was a deadly sin, they devised incredible tortures, which they inflicted on one another with a cruelty which the pagan Ammianus said could not be matched even by savage beasts, and the Christian St Gregory said was 'like hell'. But while we read† of their cutting off the ears, noses, tongues and right-hands of those who held different opinions as to whether the Son was of the same substance as the Father, or only of similar substance, we read of no one suffering for his scientific opinions. Nevertheless Christianity, with its motto 'Do not examine, only believe', must have provided a powerful deterrent to the scientific spirit of free inquiry.

In Alexandria least of all were learning and science likely to get any consideration from an all-dominating religion. Its Archbishop Theophilus, 'the perpetual enemy of peace and virtue, a bold bad man, whose hands were alternately polluted with gold and blood'‡ had a special enthusiasm for the extirpation of all monuments of pagan culture, and in 390 a large part of the great library was destroyed, it was believed, by his orders. His nephew St Cyril, who succeeded him on the archiepiscopal throne, became jealous of the influence of Hypatia; she, a pagan, was reputed to have so

* Excellent imaginary examples are to be found in Kingsley's novel *Hypatia*.

† Gibbon, *Decline and Fall of the Roman Empire*, chap. XXXVII.

‡ Gibbon, loc. cit. chap. XXVIII.

profound a knowledge of all the sciences that Christianity itself was in danger. Thus when a band of Christians, mostly monks, murdered her in 415—by tearing the flesh off her bones with sharp oyster shells—Cyril was suspected of having instigated the deed.

Some of the Alexandrians now migrated to Athens, where Plato's Academy still maintained an enfeebled existence—a small island of paganism which was gradually being submerged in the rising tide of Christianity. Although it was largely concerned with magic and superstitions, its professor of philosophy, Proclus (412–485) was the greatest philosopher of his age. He had adduced arguments against the biblical account of the Creation, and had been threatened with death, to which he made his well-known reply: 'What they do to my body does not matter; it is my spirit I shall take with me when I die.' Finally, in 529, the Christians persuaded the Emperor Justinian to forbid the study of all 'heathen learning' in Athens, and the school of Athens became dead in turn.

Others of the Alexandrians migrated to Byzantium (Constantinople), which Constantine had made his capital in 326, creating what was virtually a new city with the same zeal and thoroughness as Alexander had shown in Egypt six centuries earlier, his intention being to make a capital worthy of an Empire which was henceforth to be entirely Christian. The result was disastrous. The new city 'represented one of the least noble forms that civilisation has yet assumed'...'Immersed in sensuality and in the most frivolous pleasures, the people only emerged from their listlessness when some theological subtlety, or some rivalry in the chariot-races, stimulated them into frantic riots.'*

Learning was not likely to flourish vigorously in such a soil, but the city became a minor centre of Greek culture in the east, and remained so until the Turks captured it in 1453. If, during this 800 years, Byzantium added but little to the

* Lecky, *History of European Morals*, II, 13.

world's store of knowledge, it at least acted as a stagnant reservoir of learning from which driblets occasionally leaked out to fertilise thought outside; it created little, but it preserved much from destruction.

One of the theological conflicts of the Byzantines actually proved advantageous to learning. Nestorius, Bishop of Byzantium, maintained that the personality of Christ was a blend of two distinct natures, a human and a divine, and that the Virgin Mary was the Mother of the man Christ but not of the divine Lord Jesus; the title 'Mother of God' was abhorrent to him. When the first Council of Ephesus declared this to be a heresy in 431, the many followers of Nestorius found their lives made intolerable by persecution and moved eastward, first to Mesopotamia and thence, goaded on by more persecution, to Persia. Here they were free to occupy themselves with literature and science, writing original works in Syriac, their own native language, which had now become the common language of Western Asia, and translating the works of Aristotle, Plato, Euclid, Archimedes, Hero, Ptolemy and many others into the same tongue, with results we shall see below.

The end of the Alexandrian school came in 642, when the Mohammedans conquered the city, and destroyed the remainder of the great library. The Caliph Omar is said to have justified this final act of vandalism on the ground that 'if these writings of the Greeks agree with the book of God, they are useless, and need not be preserved; if they disagree, they are pernicious, and ought to be destroyed'. Abulpharagius records that the volumes kept the four thousand baths of the city in fuel for six months*—an obvious exaggeration, for even if there were 400,000 volumes left in the library, the average fuel ration per bath would be only four volumes a week.

* Gibbon, loc. cit. chap. XLI.

CHAPTER IV

SCIENCE IN THE DARK AGES

(642–1453)

WE have now followed the fortunes of science as it came to Europe from the east, first impinging on Ionian Greece and then penetrating to Athens, to various outlying parts of the Greek mainland and to southern Italy. Finally, when its light was already beginning to fade in Greece, it turned eastward again and found a home in Alexandria, the magnificent city which Ptolemy I had built at the mouth of the Nile.

Here many subjects of study had seemed to work themselves out to their natural endings. Geometry, which had made such magnificent progress at first, came to a dead end; algebra had hardly yet arrived; physics, which had made a good start, had been strangled almost at birth; astronomy, after making the best of starts, had taken a wrong turning at the time of Aristarchus, and was now advancing along the wrong road.

Worst of all was the opposition of religion. We have seen how the Christians had burned a large part of the great library in 390; in 415 they had murdered Hypatia; and in 642 the Mohammedans conquered the city, closed down the university and completed the destruction of the library. Each attack drove a part of the school abroad, so that learning and learned men were scattered to many lands—to Greece, to Rome, to Byzantium, even to Persia and the east. We shall now see how these scattered threads were all drawn together in the great medieval empire founded by the Arabs.

During unknown centuries Arabia had been inhabited by nomadic tribes, ranging from visionaries and dreamers to murderous savages. Their religion had been a primitive polytheism of tribal gods and devils, until Christian and Jewish ideas seeped in from Byzantium, Abyssinia and Persia.

Here in about 570 a posthumous infant Mohammed was born and brought up by a rich grandfather. He made himself a child of the desert, finally became a caravan conductor, and married an old but wealthy widow, Khadija. To her, and to his nearest relatives and friends, he confided that he had received in a vision a revelation that there was only one God, and that he, Mohammed, was his prophet. When he declared this to a wider circle, he met with ridicule and persecution, and finally he fled in 622 to Medina, where he met with more sympathy, and founded a brotherhood out of which grew the religion that was to make converts by the hundred million, and from here he preached a holy war.

The Arabs, possibly spurred on by a vision of a world-wide Mohammedan religion, now started on a career of military conquest. Palestine and Iraq fell to them within a few years; they invaded Syria in 636 and Egypt in 639; they were in possession of Alexandria in 642. Persia and western Turkestan followed, together with parts of western India, of northern Africa, of Spain and of western Europe. At breathless speed they were building one of the greatest empires the world has ever seen, but also one of the most unstable, for within four centuries its glories had departed, and it was crumbling into dust.

Their new mode of life gave them visions of a wider culture than that of the burning desert, and as they passed on their triumphal way, they absorbed learning as well as territory. Their conquest of Egypt gave them whatever of learning was left in the empty shell of Alexandria; by their conquest of Persia they acquired some of the learning which had been carried from Alexandria to Byzantium, and thence farther east by the Nestorians. Indeed there was a brief period in which the Nestorian centre of Gondisapur served as a sort of cultural capital for the mushroom empire of the Arabs, but changes soon came; Gondisapur had to yield to Baghdad, and Arabic replaced Syriac as the language of culture and science. The industrious Nestorians now set to work to retranslate the Greek classics into Arabic.

This accession of learning by conquest was supplemented by an influx from outside. Some came from Greece, carried largely by Greek physicians who were called in to treat the Arabian conquerors for a variety of diseases which had been unknown to them in their desert life. Some came from India, consisting mainly of arithmetical knowledge brought in by traders. So far the civilisation of the Hindus had contributed but little to science, possibly because the all-enveloping religious atmosphere had not been conducive to a study of material things. Life was but a passing shadow-show of which man had, for his sins, to witness many performances, and from which he must for ever try to escape through the subjugation of his personality. The material world seemed as unimportant as it had been to the early Christians, and science languished—not from persecution or intolerance, for the eastern religions ranked tolerance as a virtue—but in an atmosphere of complete unconcern. Then, as the fifth century approached its end, a tribe of Aryans invaded the country, and science began to flourish as never before or after, until the present great scientific awakening in India.

One of the more prominent Indian scientists of this early period, Arya-Batha who was born in Patna in 476, is thought to have invented algebra independently of Diophantus. He showed how to solve quadratic equations, and published a table of sines,* but we do not know whether this was his own creation or the result of having studied books by earlier writers. He also gave correct values for the sum of a series of consecutive integers ($1+2+3+ \ldots$), as well as for the sum of their squares and cubes. A later mathematician, Brahmagupta (*c.* 598–660), also solved quadratic equations and summed arithmetical progressions, but again we cannot say how far his work was original. The India of this period may not have produced much new knowledge, but it gave one great gift to the world, namely, a 'positional' notation for numbers, in which the value of a symbol was dependent on

* See footnote to p. 92.

its position—in brief our own system in which a symbol may denote units, tens, hundreds, etc., according to where it stands. Such a system was not new, for it had been used by the early Babylonians (p. 7), but it entered the western world through India and Arabia, so that we still describe our numerals as Arabic. In a later period, the Indian mathematician Bhaskara (born in 1114) wrote an astronomy which contains the first known explanation of our present-day methods of arithmetical addition, subtraction, multiplication and division.*

SCIENCE IN ISLAM

Through this combination of acquisition and influx of knowledge, the Arabians became the curators of the scientific knowledge of the world. They excelled as translators, commentators and writers of treatises, and their aim was not so much to increase knowledge as to sweep all existing knowledge into their empire. In or about the year 800 the famous Caliph Haroun-al-Raschid had the works of Aristotle and of the physicians Hippocrates and Galen translated into Arabic, while his immediate successor, al-Mamun, sent missions to Byzantium and India to find what other scientific works were suitable for translation. Conditions being as they were, the Mohammedans did no small service to science in providing a storehouse for knowledge, as the Byzantines had done before them, and assuring that knowledge which had once been gained should not be irretrievably lost.

Chemistry. In chemistry and optics, however, there is real progress to report. In chemistry two names have survived the obliterating influence of time—Jabir-ibn-Hayyan and

* The treatise was in verse and partly in the form of a dialogue with his daughter whom he never allowed out of his presence in order to prevent her marrying, e.g. 'Lovely and dear Lilavati, whose eyes are like a fawn's tell me what are the numbers resulting from 135 multiplied by 12. If thou be skilled in multiplication...tell me, auspicious damsel....' (See W. W. R. Ball, *A Short Account of the History of Mathematics*, p. 147.)

Geber. The former, who seems to have flourished in the latter half of the eighth century, explained how to prepare arsenic and antimony, how to refine metals, and how to dye cloth and leather, besides making other advances in utilitarian chemistry. He was less happy on the abstract side, introducing the fallacious idea, which was to loom large in the later story of chemistry, that matter which was burned lost something of its substance in the burning. He also added two new 'elements' to the four of the Pythagoreans and Empedocles, calling these mercury and sulphur, although he did not mean the same by these words as we mean to-day (p. 152). To these his successors added a third new element, salt.

Geber was perhaps a century later, although there is much uncertainty as to his date, some even thinking that he was the same person as Jabir.* Whoever he was, Singer† has described him as 'The father of Arabic alchemy and, through it, of modern chemistry'. Arabic alchemy, like the earlier alchemy of Alexandria, differed from modern chemistry in its aims rather than in its methods, confining itself to the single aim of transmuting substances into gold or silver. Thus we find Geber studying and improving the then standard methods of evaporation, filtration, sublimation, melting, distillation and crystallisation, as well as preparing many new chemical substances, such as the oxide and sulphide of mercury. He also knew how to prepare sulphuric and nitric acids, and the mixture 'aqua regia' in which even gold may be dissolved.

Optics. Interest was also taken in optics, and there was a growing appreciation of the possibilities of optical instruments. Legend said that the Pharos (lighthouse) at Alexandria had been equipped with some instrument through which ships could be seen at sea which were otherwise invisible; if so, no further progress seems to have been made until Arabic times. In the ninth century we find al-Kindi of Basra and Baghdad (*c*. 800–873) writing on optics, and especially on

* Dampier, *History of Science* (3rd ed., p. 79).

† *Short History of Science*, p. 132.

the refraction of light. A century and a half later Ibn-al-Haithan or al-Hazen (965–1038) was working in Cairo on the subject of refraction. He found that Ptolemy's law (p. 97) was true only for small angles, but did not discover the true law. He also studied the action of spherical and paraboloidal mirrors and the magnification produced by lenses, and solved the problem—still known as al-Hazen's problem—of finding the relation between the positions of a source of light and its image formed by a lens. He gave a correct explanation of the act of vision, saying that we see by something from the seen object passing into the eye—in opposition to the teaching of Euclid and Ptolemy that we see by something passing out of the eye and groping for the object. With al-Hazen optics was beginning to assume its modern form.

Other subjects were not entirely neglected, but there was no sensational progress. For instance, al-Khwarizmi, who was Librarian to the Caliph al-Mamun, wrote a treatise on algebra* which did much to introduce our present numerical notation into western Europe. In astronomy al-Battani, who died in 929, redetermined the constant of precession (p. 92), and calculated some new astronomical tables. At a later date Ibn-Yunas (about 1000), who was perhaps the greatest of all the Arabic astronomers, made valuable observations on solar and lunar eclipses and achieved substantial progress in trigonometry.

But the age was less remarkable for its scientific advances than for its succession of men of encyclopaedic knowledge, each writing on a vast variety of subjects. al-Kindi, 'the first philosopher of the Arabs', whom we have already mentioned,

* *Al-gebr we'l mukabala.* The first word of the title, from which our word algebra is derived, means restoration, and refers to the transfer of a quantity from one side of an equation to the other by the process of adding the same quantity to both sides of the equation or subtracting the same quantity from both sides.

Algebra is one of the few exceptions to the general rule that the sciences take their names from the Greek language—arithmetic, geometry, trigonometry, physics, astronomy, and so on.

issued 265 publications on the most varied subjects, while the Persian Rhazes (865–925), who was primarily a physician and an exceedingly good one, not only wrote on measles and small-pox, but also on alchemy, theology, philosophy, mathematics and astronomy. There was also al-Biruni (973–1048), who was mathematician and astronomer, physicist and geographer, physician and historian. It was in the last of these capacities that he achieved the greatest fame, but he also determined the specific gravities of a number of metals and precious stones by the method of Archimedes.

Mohammedan science flourished in a subdued way until about the end of the tenth century, and then conditions began to change. The golden age of Islam had already passed away, and now the great Empire was itself breaking up, its ruling classes dying out and its more distant provinces seceding. Culture was on the downgrade and science with it. In the east, at least, it had definitely outstayed its welcome, and was coming under attack as being antagonistic to religion and causing 'a loss of belief in the origin of the world and its Creator'. The Mohammedans of the east were soon as unsympathetic to science as the Christians had been before them.

As Mohammedan science wilted in the east, it acquired a new vitality in the west, beginning in Spain, and more especially in Cordoba and Toledo. In Cordoba an academy and library were established under the special encouragement of its Caliphs, Abd-ar-Rahman III and al-Hakam II. Gradually an interest in Arabic ideas and an appreciation of Arabic learning spread over western Europe. We find Gerbert, who was afterwards Pope Sylvester II and died in 1003, introducing an Arabian form of the old Roman abacus, while another ecclesiastic, Herman the Cripple (1013–54), of the monastery of Reichenau in Switzerland, wrote books on mathematics and astrology which showed a strong Arabic influence. An Englishman, Adelard of Bath (about 1090–1150), who had disguised himself as a Mohammedan student and attended lectures at Cordoba, wrote a compendium of Arabic science under the title *Natural Questions*, while Arabic alchemy was introduced to the western world in 1144 by another Englishman, Robert of Chester (about 1110–60), who lived in Spain for many years and finally settled in London in 1147. Somewhat later, another Englishman, the Yorkshireman John de Holywood (Latinised as Sacrobosco), wrote an *Astronomy* which contained little beyond translations from Arabic writers, but remained the standard text-book on the subject for some time.

At the same time, a whole flood of classical books was being translated from the Arabic into Latin, so that the works of Aristotle, Euclid, Archimedes, Appollonius and others became available to the cultivated world in a language they could understand. Adelard of Bath had secured a copy of Euclid's *Elements* in Arabic during his sojourn in Cordoba, and made a translation which formed the basis of all European editions of Euclid until the original Greek text was recovered in 1533. Shortly after this, the Spaniard Domenico Gonzales of Toledo translated the physics and other works of Aristotle into Latin, while John of Seville did the same for the astronomical and astrological writings of al-Battani, al-Khwarizmi, al-Farabi, al-Kindi and others. But surely the most industrious translator must have been Gerard of Cremona (1114–87), who learned Arabic through a sojourn in Toledo, and is said then to have translated ninety-two complete works from Arabic into Latin, including Ptolemy's *Almagest*, Euclid's *Elements*, and works by Apollonius, Archimedes, al-Battani, al-Farabi, Geber and al-Hazen.

Besides these endless translations, the Spain of this period produced a small amount of original thought, especially in astronomy. The astronomer Arzachel, a Cordoban who lived in Toledo in about 1080, anticipated Kepler (p. 164) by suggesting that the planets moved round the sun in ellipses, but found that no one was willing to consider a hypothesis that was so opposed to the doctrines of the *Almagest*. About a century later al-Bitrugi of Seville (Alpetragius in Latin) proposed replacing the complicated Ptolemaic system of cycles and epicycles by a system of concentric circles. When his book was translated into Latin by Michael the Scot (about 1175–1235), it carried the first challenge to the Ptolemaic astronomy into western Europe.

One of the last gifts which Mahommedan science transmitted to the western world was the 'Arabic' system of numbers, which the Arabs had themselves acquired from India (p. 105). Adelard of Bath had first introduced it when he translated al-Khwarizmi's *Arithmetic* into Latin early in the twelfth century, but a more conscious effort was made by the much-travelled Italian mathematician Leonardo of Pisa, when he asserted in his best-known book* that the system was but little known in Europe, and recommended it as being more convenient than the commonly used Roman system. Shortly after this, John de Holywood used the system in a much read text-book on arithmetic, which, like his astronomy, remained the standard text-book on the subject for a long time. A few years later, in 1252, King Alphonso (the 'Wise') of Castille had some Toledo Jews compute new astronomical tables from Arabic observations and publish them in the Arabic notation. Through these and other similar activities, the Arabic notation gradually became understood, and was in fairly common use by the end of the thirteenth century. At this same period we come to the end of the era of translations and text-books, in which so many had tried to recapture the knowledge of former ages, and so few to extend it. Science had now come back to the west, and was free to advance by western methods.

If we try to sum up the advantages which had accrued to science from its stay in Islam, we think first of its new notation for numbers, and its new methods of manipulating numbers. For the rest, a knowledge of algebra had been gained which was almost identical with our present knowledge of elementary algebra. Geometry still stood where it had at

* *Algebra et Almuchabala*, or *Liber Abaci* (1202).

the end of the Greek supremacy, but there was little need for it to advance now, since algebra and trigonometry could do all that was wanted. Physics had freed itself from the speculative atmosphere which had enveloped it in Greek times, and had become experimental instead of contemplative—an enormous step in the right direction. To determine the specific gravities of precious stones by a thousand-year-old method would seem a dull piece of research to a modern physicist, but it lies on the high road to his present point of vantage, whereas the Greek orgy of speculation could lead nowhere. Science had also gained a new appreciation of the value of optical instruments, although we do not yet hear of any attempt to use them for astronomical purposes. Chemistry, too, had made a start along the right road, but was not yet entirely disentangled from a fraudulent alchemy.

WESTERN SCIENCE

It must not be supposed that while science had been making these gains in Islam, it had been completely stagnant elsewhere. It had not, but it had enjoyed only spasmodic vitality, flourishing at the best only in isolated and transitory patches. The story of such a spasmodic period of activity usually began with a stirring from the top, frequently by a very highly placed personage, which failed to evoke any real interest in the masses of the population, few of whom had the education necessary for an interest in science. Any interest there was in genuine science was usually transferred in the end to the spurious sciences of alchemy, astrology and magic; these could claim to be advantageous to their devotees, while real science, offering only knowledge for its own sake, could make no such claims.

A conspicuous instance occurred in 787, when Charles the Great resolved to encourage learning in his empire, and decreed that every abbey must establish a school. He charged two monks, Peter of Pisa and Alcuin of York, who were attached to his court, with the carrying out of the order, and through their efforts some learning was brought back from east to west, but many centuries were to elapse before there was any widespread interest in science. Similarly, in the tenth century two of the Byzantine Emperors, Leo VI and Constantine VII, showed an enthusiasm for astronomy, but little of this spread even to the educated layers of the population.

We can hardly leave the question of royal interest in medieval science without pausing for a moment to glance at the fantastic and arresting figure of Frederick II, Emperor of the Holy Roman Empire (1194–1250), whom his friends called 'Stupor mundi'—the world's wonder. He was so talented and versatile, whether as scholar and poet, or as soldier and statesman, or even as a mere linguist, that the world could in no case have overlooked him. But he was careful not to let it, and was assiduous in drawing the maximum of attention to himself, keeping up a great harem and travelling with an assortment of elephants, dromedaries and other arresting animals which could hardly fail to be seen even in the colourful thirteenth century.* He was said to have denounced Christ, Moses and Mohammed as a trio of imposters, and engaged in a series of quarrels with the Pope, who excommunicated him twice—first when he failed to start on a crusade which he had vowed to undertake, and secondly when he decided to go after all. Yet his hectic and vivid personality found time and energy for a genuine interest in the things of the intellect—philosophy and mathematics, astrology and medicine in particular—and he showed his interest by active help. It was the age when the great medieval universities were coming into existence,† and Frederick was personally responsible for the foundation of Naples and Padua. He further had a number of Arabic works translated by a band of Jews. It is not clear whether his primary aim

* H. A. L. Fisher, *A History of Europe*, p. 272.

† The dates of recognition by the state were: Paris, 1200; Oxford, 1214; Naples, 1224; Cambridge, 1231; Padua, 1222.

was to help science or to annoy the Pope—he succeeded in doing both—but the results were good, and copies of the works of Euclid, Archimedes, Apollonius, Ptolemy and others were made available by his action.

On one occasion he broke a journey at Pisa to test for himself the far-famed mathematical powers of Leonardo of Pisa, and arranged a mathematical tournament at which a 'problem paper' was set to all comers. This has been preserved, and the problems are of interest as showing the high mathematical level of the period. One problem (stated in modern language) was to find a number x such that x^2+5, x^2 and x^2-5 are all squares. Leonardo gave the correct solution, $x=\frac{41}{12}$.* Another problem was to solve the equation $x^3+2x^2+10x=20$ by geometrical methods. Leonardo showed that this is impossible, but gave the algebraic solution $x=1\cdot3688081075$, which is accurate to nine places of decimals.† Leonardo came out quite triumphantly, solving many of the problems correctly, while no one else solved any at all.

Science in the Monastic Orders

Not only were the medieval universities founded at this time, but also two monastic orders—the Franciscans or Gray Friars in 1209, and the Dominicans or Black Friars in 1215—both of which had their influence on the progress of science. At first the events were of purely religious significance. St Francis, the son of a rich merchant of Assisi, experiencing a sudden

* The mathematician will see that this is far from mere child's play. If x is the solution, then x^2+5 and x^2-5 must be of the forms $x^2+5=(x+y)^2$, $x^2-5=(x-z)^2$. Eliminating x from these two equations, we obtain

$$\frac{5}{z+y}=\frac{yz}{z-y},$$

and each fraction is equal to $\dfrac{\sqrt{[5^2-(yz)^2]}}{2\sqrt{[yz]}}$. Presumably x, y and z must all be commensurable, which condition is satisfied if $5^2-(yz)^2$ and yz are both perfect squares. An obvious solution is $yz=4$, which leads at once to Leonardo's value for x.

† Leonardo still used the sexagesimal scale of the early Babylonians, and so expressed his solution in the form $x=1\ 22'\ 7''\ 42'''\ 33''''\ 4^{\text{v}}\ 40^{\text{vi}}$.

conversion, abandoned a gay, careless and wild life to devote himself to the relief of suffering and the preaching of repentance. This he did with enthusiasm; we read of his jumping down from his horse to kiss a leper that he met on the road, and preaching the glad tidings of the gospel to the birds and fishes. He founded an order of friars who were at first intended to follow in his footsteps, preaching to simple people in simple language. But they soon found that the land was infested with heresies of all kinds, and they devoted themselves to the acquiring of learning, so as to be able to refute them.

The Dominicans were cast in a different mould. Their founder, St Dominic (1170–1221) was a professional theologian who had already attained to cathedral dignity when he founded his order. Stern and austere in his personal life, he burned with zeal for the extirpation of every kind of heresy, but most particularly for that of the Albigensians, who maintained that there were two Gods, one good and one bad, the fathers respectively of Jesus and Satan. After himself preaching against this heresy for ten years, St Dominic founded his order of Preaching Friars who, living in extreme poverty and asceticism, were to spread true doctrines throughout the world. They, too, found that their missionising called for a fund of knowledge. They made a special effort to gain a footing in the universities, and held chairs in most of them, while their special ardour for orthodoxy resulted in a fanaticism and intolerance which found its obvious outlet in the Inquisition, so that the Chief Inquisitor in most countries was a Dominican.

The members of these two orders provided a good proportion of the scientists and teachers of the next two centuries, the scientists coming mainly from the Franciscans, while the Dominicans produced others who figured prominently in the history of thought.

ST THOMAS AQUINAS. Foremost among the latter was of course St Thomas Aquinas, the greatest of all medieval theologians. In his *Summa contra Gentiles* (1259–64) he argues that knowledge can be obtained through the two distinct channels of faith and natural reason. Faith derives its knowledge from Holy Scripture, natural reason from sense-data which it interprets and transforms by a process of ratiocination, of which the works of Plato and Aristotle provide the supreme example. Since both kinds of knowledge come from God, who cannot contradict Himself, they must be in agreement. It follows that the writings of Plato and Aristotle must be agreeable with the doctrines of the Christian religion, and in his *Summa Theologica* St Thomas considered he had established that this is actually so. Building on this basis, he developed the philosophical system now known as 'scholasticism'. It takes its name from the system of schools which Charles the Great had instituted in the eighth century (p. 111), but it amounted to little until St Thomas developed it into a consistent body of doctrine in the thirteenth century. This century saw its rapid rise, but the two succeeding centuries saw its equally rapid decline and fall. By giving itself over to logic-chopping, to abstract subtleties and trivialities which were of no interest to live men, it soon lost its hold over thinking humanity, and by the sixteenth century it was dead, blown out of existence by the fresh breezes of the renaissance. As it rose and fell, so also did faith in the infallibility of Aristotle.

The more gentle and more human Franciscans were less concerned with the accuracy of their opinions, and more with the accuracy of their knowledge, which they tried to check by a direct comparison with the works of God. Prominent among the scientists of their order were many who held high positions in the Church. Robert Grosseteste (*c.* 1175–1253), Chancellor of Oxford University and Bishop of Lincoln, and John of Peckham (*c.* 1220–92), were two such. Both wrote on optical problems of the kind discussed by al-Hazen, while Grosseteste himself performed experiments with mirrors.

ROGER BACON (*c.* 1214–94). But the most important of the Franciscan scientists was a simple friar who attained to no high office, either in the Church or outside it. He was born near Ilchester in Somerset, and studied first at Oxford and then in Paris. Little is known for certain about his subsequent life, but he seems to have returned to Oxford somewhere about 1250, and lectured there with great success. Nevertheless his wealth, which had been substantial, soon melted away, and he took the vows of the Franciscan order, only to find that a purely religious life contained no satisfaction for him, while an attempted return to scientific activity incurred the displeasure of his monastic superiors. For ten years he was kept under restraint and forbidden to write. Then in 1266, to his unspeakable joy, his old acquaintance Guy de Foulques, now Pope Clement IV, invited him to resume his scientific work, and is thought to have pleaded his case personally with the Franciscan authorities. Permission was finally given, and within two years Bacon had dispatched to the Pope his *Opus Majus*, which was a sort of general compendium of the scientific ideas and knowledge of the period. But Clement died in 1268 and Bacon was soon in renewed trouble with his Franciscan superiors; in 1278 he was tried at Paris, condemned for unorthodox opinions, and spent most of his remaining life in imprisonment.

Rumour said that Bacon was not only interested in true science but also in the black arts; indeed, it is in the character of a lurid necromancer that he was known to the world at large. In science his main interest was in optics. He understood the laws of reflection and refraction of light, and explained how lenses could be arranged to act as spectacles (the invention of which is frequently attributed to him) and telescopes, although there is no record of his ever having made either himself. But this was by no means his only interest, and we find his mind ranging over most of science in an imaginative and fanciful, although often not very practical, way. He described how mechanically propelled carriages, ships and flying machines might be constructed—the imaginary forerunners of our motor-cars, steamers and aeroplanes—and discussed possible uses of gunpowder and burning-glasses, the circumnavigation of the globe, and other things which seemed strange in his time, but have become commonplaces to-day. He argued against the 'naturalness' of circular motion, condemned the Ptolemaic system of astronomy as unscientific, and thought it was probably untrue.

But his general principles were more important than his detailed achievements, which were after all meagre enough. In his *Opus Majus* he argued that mathematics should be at the foundation of all liberal education, since it alone 'can purge the intellect and fit the student for the acquirement of all knowledge'. He also insisted that scientific knowledge could only be acquired by experiment; this alone gave certainty, while all else was conjecture.

This seems obvious enough to-day, but it was not so when Bacon wrote. There was still but little idea of taking nature's verdict, as revealed by experiment, as the final arbiter of truth. Men were skilled neither at asking questions of nature nor at interpreting her answers. They might indeed ask whether an alleged fact was in agreement with experiment, but they would first ask (and it was an easier question to answer) whether it was in agreement with Aristotle or conformable with Holy Scripture. And those who held that Aristotle and Scripture, as representing reason and revelation, must necessarily agree with one another and with the truth, did not usually go so far as to inquire as to the findings of experiment.

Bacon objected to this last frame of mind, but went no further; he was no revolutionary, but a child of his age, and was as firmly convinced as his contemporaries that in the end science must be found to agree with the Christian religion, and so establish its truth. He came at the end of an epoch, but it was his death—not his birth or life—that marked this end. A new era was dawning in which men would no longer try to discover the truth by reading the opinions of ancient writers, but from a first-hand examination of the works of God.

Signs of the coming Dawn

It is not altogether easy to see why this last change occurred; it is certainly much too superficial to explain it by the single word 'renaissance'. The classical renaissance in literature had hardly yet acquired any strength, and what influence it had was in the direction of turning men's thoughts back to the ideas of the ancients. In science the movement was in the contrary direction. Science was profiting vastly by its access to Greek scientific writings, but the trend of thought was away from Greek scientific methods. Perhaps the explanation is that scientific knowledge, unlike literary imagination, is cumulative, and that medieval science had already reached a stage in which it had more to offer than Greek science, the reverse of what was happening in the field of literature.

The first science to benefit from the new independence of thought was astronomy. The first name to attract our attention is Oresme, Bishop of Lisieux (1323–82), a man of great and varied abilities, who had been confidential adviser to Charles V of France, and then tutor to Charles VI. He was not only a prominent ecclesiastic and theologian, but was also distinguished in mathematics and economics. He wrote a treatise on currency which is noteworthy for its use of vulgar fractions like those we use to-day—$\frac{3}{4}$, $\frac{1}{8}$, etc. But his main interest to us is that he challenged the Aristotelian doctrine of the fixity of the earth.

Nearly a century later his challenge was repeated by Nicolas of Cusa, a fisherman's son who had risen without influence to be a Cardinal of the Church. Entirely rejecting the traditional astronomy, he expressed the opinion that the earth 'moves as do the other stars'.

The last five names we have had occasion to mention have all been those of ecclesiastics of one sort or another—education

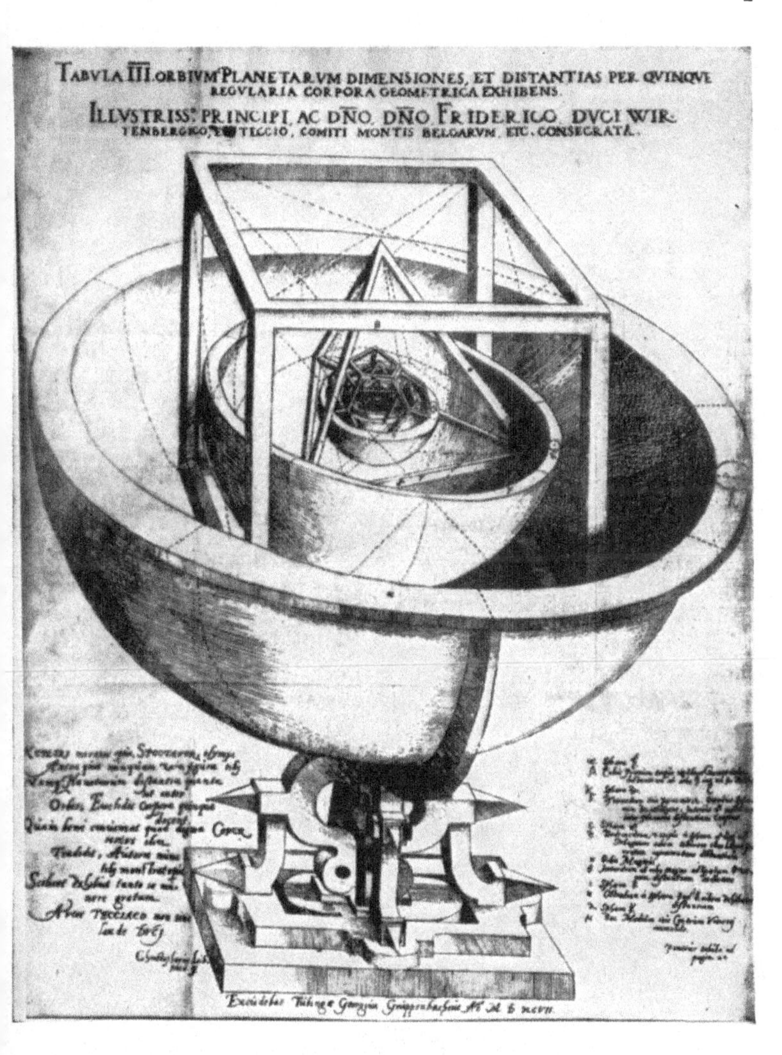

Orbits of the planets

Illustration from Kepler's *Mysterium Cosmographicum*, 1596, showing the orbits of the planets filled between the octahedron, icosahedron, dodecahedron, tetrahedron, and cube

was almost the exclusive prerogative of the Church. One of the five, Roger Bacon, was in almost continuous trouble for his scientific activities (but never, so far as we know, for his scientific opinions), while the other four, who all held high positions in the Church, seem to have been allowed not only to study science, but also to express opinions which were opposed to the traditional teachings of the Church. It was a flat denial of the scriptural account of creation to teach that the world was spherical, yet this was commonly taught, and the new view of the earth as a moving star seems to have been regarded in the same way. Broadly speaking, the Church of the moment was tolerant to the further progress of science, soothed perhaps by the common belief that science was bound in the long run to support and confirm orthodoxy; the Church could, so to speak, afford to be tolerant, at least for a time.

Many other factors combined with this to presage a bright future for science. Not only was the human mind regaining its long-lost freedom of thought, but it now had at its disposal the almost endless writings of the greatest age of ancient thought, as well as of later ages. One thing more was needed and, just when it was most needed, it came.

The earliest scientific knowledge had been spread by word of mouth. Afterwards came the age of the great libraries, such as those of Alexandria and Byzantium. Inside these many thousands of books could be read by those who had the means and leisure to travel to them. Outside them, books were rare and costly articles, since the production of a single copy involved a colossal task of writing or transcription on costly parchment—perhaps ten times as much labour as is expended to-day in setting the type for an edition of thousands of copies.

The Chinese had invented a kind of paper early in the Christian era, which they had used for printing from blocks in the ninth century, and for printing from movable type in the eleventh. In the fourteenth century this latter art was discovered independently in Europe, and from this time on printed books became available in ever-increasing quantities, until now every man is able to have a library of his own in a few feet of shelving by his fireside.

In time, but not at once, the coming of printing made scientific knowledge more accessible, and so more widespread. Religious and literary works were thought to have the first claim on the newly established presses—first the Bible (1456) and then Greek and classical writers. No scientific work was printed until Pliny's *Natural History* appeared in Venice in 1469. This was followed in 1471 by the works of Varro, a Roman country gentleman (116–27 B.C.), who had written an encyclopaedia of the sciences. So far the choice of scientific authors had not been particularly happy, but now there was an improvement. A Latin translation of Ptolemy's *Geographia* appeared in 1475, and three biological works of Aristotle, also in Latin, in 1476. Euclid was published in Latin in 1482, and an adequate edition of Aristotle in Greek in 1495. But Ptolemy's *Almagest* was not printed until 1515, and Archimedes' *Psammites* only in 1544, the former being issued at Venice and the latter at Basle.

Everything was now favourable to a period of scientific activity, and it came—little rivulets in the sixteenth century, and an overwhelming torrent in the seventeenth.

CHAPTER V

THE BIRTH OF MODERN SCIENCE

(1452–1600)

THERE have been many attempts to assign a precise date to the beginnings of the literary renaissance—as, for instance, 1453, the year in which Byzantium was captured by the Turks, and the treasures of its library scattered throughout Europe. But it is a futile task; the renaissance did not come overnight; but was a slow development of centuries.

It is the same with the renaissance of the scientific spirit, with which we shall be concerned in the present chapter. This revived only gradually after its thousand-year torpor. But if we had to select a single year, a good deal could be said for 1452, the year preceding that just mentioned. For in this year was born Leonardo da Vinci, whom many hail as having been the first scientist to disentangle his thought from all the confused and erroneous ideas of the Middle Ages, and to approach the study of nature in a truly modern spirit. With Leonardo, science adopts modern aims and modern methods. Thus it is not inappropriate to begin the present chapter with a brief mention of this truly extraordinary man.

LEONARDO DA VINCI (1452–1519). His birthplace lies near Empoli on the road from Florence to Pisa. The natural son of a Florentine lawyer, and of a common peasant girl who afterwards married a cowman, his fine appearance and engaging manner marked him out as one obviously suited to court life, and actually he was associated with the courts of Florence, Milan and Rome in turn. But his bodily gifts, outstanding though they might be, were not to be compared with those of his mind. Students of his work have long credited him with almost superhuman intellectual powers; his note-books, recently recovered and deciphered, confirm these estimates and even add to his reputation as one of the outstanding intellects of the human race.

He was primarily an artist, devoting his main energies to painting and sculpture, but he achieved distinction in a number of other fields as well—in architecture and engineering, in philosophy and science—and the note-books suggest that he could have done the same in yet others had he cared to do so. He had planned to write text-books on all the various subjects of his study, and if only these had materialised, science might have been saved from taking many wrong turnings.

Unhappily his defects were almost as outstanding as his talents. Outside his chosen field of art he seems to have worked clumsily, laboriously and slowly, so that his output of finished scientific work was minute in amount—indeed, he seldom finished anything. His best-known positive achievement is probably his explanation of the dim illumination which appears on the dark part of the moon at times of new moon—'the old moon in the young moon's arms'. Leonardo rightly attributed this to 'earthshine'—sunlight reflected from the earth. He also accomplished some experimental work of a very practical kind in optics, mechanics and hydraulics. In the more utilitarian sciences he made plans and designed models for flying machines, helicopters and parachutes, as well as for quick-firing and breech-loading guns. His 750 anatomical drawings put him in the front rank of the anatomists of the world.

But it was his unproved speculations and his unverified opinions that showed his scientific talent most markedly. In physiology he anticipated Harvey's discovery of the circulation of the blood, conjecturing that the flow of blood in the human body is like the cyclic flow of rain water which falls on the hills as rain, flows thence to the rivers and seas, is carried back to the clouds, and then completes the circuit by coming down again in the form of rain. He said that blood brings new matter to the various parts of the body and

A row of four mortars firing into the courtyard of a fort

Sketch by Leonardo da Vinci. *c.* 1503–4

carries off the waste products, much as we stoke a furnace and carry the ashes away. In astronomy he thought that the earth was 'a star like other stars' (meaning a planet like other planets), and he hinted at a heliocentric world, although others had of course done this before him. In mechanics he said that 'every body has weight in the direction in which it is moving', and asserted that a falling body increases its speed as the fall progresses. Thus he seems to have understood that force primarily produces acceleration rather than mere movement, challenging the Aristotelian doctrine that force is necessary for movement, and anticipating the characteristic feature of Galileo's mechanics (p. 145). He suggested that the whole universe conforms to unalterable mechanical laws, which was of course only a repetition of earlier speculations of Democritus and Anaximander, but it also anticipated Newton, although without demonstration or proof. In optics he regarded light as an undulatory phenomenon, thus repeating many earlier suggestions, but also anticipating the undulatory theory of light.

All this was guesswork, but it was the guesswork of a mind thinking freely and unshackled by authority, and it must be conceded that it shows a certain quality of genius. Good luck may enable any fool to make a sound guess or a true prediction here and there, but when hit after hit is scored in almost unbroken succession, something more than luck must be at work. To form some idea of the value of Leonardo's conjectures, let us pause to think what the gain would have been if his opinions could have replaced those of Aristotle as the touchstone against which every theory had to be tested before it could command the attention of the learned world.

Perhaps Leonardo's greatest service to science was his exposition of the principles which ought to govern scientific research. Before his time men had often enough tried to advance knowledge by observation and experiment, but their ideas had too often been cramped by preconceived ideas or general principles such as the naturalness of circular motion or suitability to some imagined purpose. Leonardo was very cautious about invoking general principles. It is true that he worked out the mechanics of the lever from the principle that perpetual motion is an impossibility, but this principle does not embody a preconceived notion of how things ought to be, but thousands of years of experience as to how they actually are, experience which is none the less real for having been acquired largely unconsciously.

Leonardo followed Aristotle in insisting that mathematical reasoning alone can give complete certainty in science, but he parted company from Aristotle in seeing that this is a counsel of perfection; in most sciences certainty is an unattainable ideal. A science, Leonardo says, should be based on observation; it may properly use mathematics to discuss the observations, and ought preferably to end with a crucial experiment to test its final conclusions. His general views on scientific method were very like those which Roger Bacon had expressed a century earlier, but Bacon's vision had ever been restricted by theological blinkers, whereas Leonardo's mind worked perfectly freely.

This new outlook on the methods of science was followed by a new outlook on the universe and man's position therein. Copernicus brought forward evidence that Aristarchus had been right in maintaining that the earth did not hold the central position in the universe but, like the other planets, was a mere wanderer in space which described a yearly orbit round the sun.

ASTRONOMY

COPERNICUS (1473–1543). Mikolaj Koppernigk (latinised as Nicolaus Copernicus) was born at Torun (Thorn) in Polish Pomerania on 14 February 1473. His father, an eminent citizen of the town, had been born in Cracow, to which his family had previously migrated from Silesia. There is more uncertainty about the ancestry of his mother, but it is thought that she came of a wealthy Silesian family. When Nicolas was 10 years old, his father died, and his uncle, an eminent ecclesiastic, educated him for high office in the Church. After leaving school he studied at various universities until he was more than 30 years old—first at the Polish University of Cracow, and then at the Italian Universities of Bologna, Ferrara and Padua in turn.

It was a prolonged education, but a cultivated man of those days regarded all knowledge as his province, and did not limit his education to acquiring skill in one particular calling. In this spirit Copernicus acquired a profusion of knowledge in the classics, in mathematics and astronomy, medicine, law and economics, and of course in theology. This he did not let lie idle, but put it to full use even after he had reached a high position in the Church; we read of his curing the sick poor as well as sick fellow-ecclesiastics, writing on economics, and advising the Polish government on questions of currency,* making his own scientific instruments, writing poetry and even painting, at least to the extent of one self-portrait. He also achieved success as an administrator, as an estate manager and as a diplomat at a minor peace conference. Like Leonardo before him, he was a man of wide knowledge and of varied attainments, but his main interest was ever in mathematics and astronomy.

These subjects figured largely in the teaching of the medieval universities, and we find Copernicus attending lectures on Euclid, on spherical geometry, on geography, on astrology and on Ptolemaic astronomy.

This last was still the official astronomy of the universities and of the Church, but a number of advanced thinkers were already feeling sceptical about it, and advocating something more like the heliocentric astronomy of Aristarchus; we have already noticed the astronomical opinions of Bishop Oresme of Lisieux, Cardinal Nicolas of Cusa, and Leonardo da Vinci.

* See p. 186 below.

Another of the same way of thinking was Domenico Novaro, who was Professor of Mathematics and Astronomy at Bologna while Copernicus studied there and remained his friend after he had left. The writings of the Pythagoreans, which had now become accessible to European scholars, proclaimed that the ultimate truth about the universe must consist of simple, elegant and harmonious relations, and Novaro thought the Ptolemaic astronomy too cumbersome to comply with this criterion. We may assume that his doubts and criticisms influenced the thoughts of the younger man, so that when the latter returned to Poland to a Canonry in the Cathedral of Frauenburg, he took the problem of the Ptolemaic astronomy with him.

His reading showed that the philosophers of antiquity had held varying opinions as to whether the earth stood at rest or was in motion. In the dedication of his magnum opus *De Revolutionibus Orbium coelestium* to Pope Paul III, he recalls that 'according to Cicero, Hicetas had thought that the earth moved...and according to Plutarch others had held the same opinion'.* This, said Copernicus, led him to long meditations on the subject, which ultimately resulted in the system proposed in his book.

Copernicus begins his discussion by remarking that 'every change of position which is observed is due to the motion either of the observed object or of the observer, or to motions of both....If the earth should possess any motion, this would be noticeable in everything that is situated outside the earth, but would be in the opposite direction, just as though everything were travelling past the earth. The relation is similar to that of which Aeneas says in Virgil: 'We sail out of the harbour, and the countries and cities recede.' Copernicus

* In his *De Placitis Philosophorum* Plutarch writes: 'Aristarchus places the sun amongst the fixed stars and considers that the earth moves round the sun.' The first authentic translation of this work appeared in a book by Georg Valla, *De expetendis et fugiendis Rebus* (1501), from which Copernicus borrowed a diagram and an almost verbatim description of the views of Aristotle and Aristarchus on the vastness of the universe.

next suggests that the apparent daily revolution of the 'sphere of the fixed stars' can be explained on these lines, the earth, and not this sphere of stars, rotating once a day. This opinion, he says, was held by the Pythagoreans Heraclides and Ecphantus, and by the Syracusan Nicetas (as told by Cicero), who assumed the earth to be rotating in the centre of the universe. He continues: 'It would thus not be strange if someone should ascribe to the earth, in addition to its daily rotation, another motion also. It is said that the Pythagorean Philolaus, no ordinary mathematician, believed that the earth rotates, that it moves along in space with various motions, and that it belongs to the planets; wherefore Plato did not delay travelling to Italy to interview him.'

We have seen how the Ptolemaic system placed the earth at the centre of the universe and supposed the sun to move round it in a circular orbit. Out beyond the orbit of the sun lay three more circular orbits in which nothing moved but mathematical abstractions known as fictitious planets. The real planets Mars, Jupiter and Saturn moved round these fictitious planets in small circles known as 'epicycles'. These were all of the same size, and at every instant the planets were at 'corresponding' positions in their epicycles, i.e. the lines drawn from the fictitious planets to the real planets all pointed in the same direction, which was the direction of a line drawn from the earth to the sun (see fig. 20, p. 95).

This was obviously a very artificial arrangement, but its very artificiality provided a clue to its true meaning. When a child sits in a whirling 'merry-go-round' at a fair, distant objects and spectators seem to the child to advance and recede alternately—as a result of course of the child's own motion. If the child moves in a circle of 20 ft. radius, then outside objects will appear to the child to move in circles of 20 ft. radius, and at every moment they will all appear to be in corresponding positions in these circles. The apparent motion of many objects is a sort of 'reflection' of one real motion, namely, that of the child.

Copernicus thought that the apparent motions of the planets in their epicycles could be explained on similar lines as reflections of one real motion of the earth round the sun. If so the whole motion of the solar system was one of the earth and planets moving in circular orbits round a fixed central sun, so that the earth was simply one 'wanderer' of many, while 'the sun, as if sitting on a royal throne, governs the family of stars which move round it'. It was not a new hypothesis, being identical with that which Aristarchus had proposed some 1800 years earlier. What Copernicus had done was to show that the old system of Aristarchus could explain the observed motions of the planets, or rather would produce precisely the same appearances in the sky as the complicated epicyclic motions of Ptolemy.

But this was true only of the Ptolemaic system in its simplest and most primitive form, and it had long been known that this did not fit the observed motions of the planets with any great accuracy; it had needed amendment time after time, first by Ptolemy himself and later by his Arabian successors, until it had become very complicated indeed. Copernicus could neither be sure of his own scheme nor expect others to accept it, unless he could amend it to fit the best observations available. It was here that Copernicus found his most serious task, to which he devoted years of arduous toil.

Most of it was wasted. We know now that only one small amendment was needed, namely, replacing the circular orbits of the planets by slightly oval curves—to be precise, by ellipses (p. 40) which were nearly, but not quite circular; Kepler showed this in 1609. But Copernicus, with his mind still soaked in Pythagorean and Aristotelian doctrines as to the 'naturalness' and 'inevitability' of circular motion, could not think of anything so simple as scrapping the circles of the Ptolemaic scheme; the most drastic amendment to his new scheme that he could imagine was an addition of more circles. So he added some new circles in the form of epicycles similar to those he had just discarded, and increased the complexity of the circles already in the scheme by making these 'eccentric', i.e. by supposing that the centres of the planetary orbits did not coincide with the sun. In brief, he tried to get a better agreement with the observations by just the same threadbare devices as had been employed by the mathematicians of the last 1400 years.

His inexperience as a practical astronomer increased his difficulties, for he paid equal attention to all observations, good and bad, ancient and modern, including a few of his own, so that one bad observation could throw the whole system out of gear. Actually he twice introduced unnecessary complications to make room for phenomena which, as we now know, had no existence except in his own faulty observations. He did not aspire to a very high level of accuracy, for he is reported to have said that if he could get an agreement to within 10′ of arc he would be as elated as Pythagoras had been at the discovery of his famous theorem.

At last, after years of arduous toil, the system stood complete. It naturally passed all observational tests, for he had introduced one complication after another to make it do so. These made its details as complicated as its central idea was simple. The Ptolemaic system had needed about eighty circles to explain the known phenomena; Copernicus still had to employ thirty-four. The complexity had been alleviated rather than cured.

Copernicus sent a brief summary of his conclusions to his astronomical friends under the title *Commentariolus*, but he shrank from the task of preparing the whole work for publication, so that ten years elapsed before he consented to do this. But he gave permission to a certain Georg Joachim (Rheticus), who had resigned from a chair of mathematics at Wittenberg to go and work with him, to prepare a short summary of the contents of the book, and this was published in 1541 under the title *De Libris Revolutionum Narratio prima*. Copernicus subsequently handed over the text of his whole book to Rheticus, who undertook to revise it and prepare it for the press. It is said that the printing was completed only just in time for the author to handle the book as he lay on his death-bed.

In his dedicatory letter to the Pope, Copernicus explained why he had hesitated so long before publishing his theories to the world: 'I considered what an absurd fairy-tale people would consider it, if I asserted that the earth moved....The scorn which was to be feared on account of the novelty and absurdity of the opinion impelled me for that reason to set aside entirely the book I had already drawn up.'

Notwithstanding this clear statement, it has often been asserted that Copernicus held back his book from a fear that it might incur the displeasure of the Church. It is hard to find any evidence for this. Copernicus had made no secret of his conclusions, but had circulated his *Commentariolus* to many high officials of the Church; many had urged him to publish his 'revelation of the truth', and he had been given permission to publish his *Narratio* to a wider circle in 1540. Actually we hear of no serious ecclesiastical opposition to the book until 1616, when it was put on the 'Index', so that all good Catholics were forbidden to read it. But many things had combined to produce a different atmosphere by 1616.

The Lutherans, however, from Luther and Melanchthon down, had hated the book from the first, and fired many broadsides of bad argument against it.* Possibly they were more keen-sighted than the official Church in seeing how unorthodox its religious implications were.

Rheticus had not seen the book through the press himself, but had entrusted the final stages to a friend of Copernicus, a Lutheran minister named Andreas Osiander, who had made no secret of his fears that the book might offend the Lutherans,

* Melanchthon argued: 'When a circle revolves, its centre remains unmoved; but the earth is at the centre of the world, therefore it is unmoved.' Luther wrote: 'It says in Holy Writ that Joshua bid the sun stand still, not the earth.' See H. Kesten, *Copernicus and his World*, London, 1945.

and had suggested describing its conclusions as only hypothetical. He had written to Copernicus: 'For my part, I have always felt that they [hypotheses] are not articles of faith but bases of calculation, so that even if they be false, it matters not so long as they exactly represent the phenomena....It would therefore seem an excellent thing for you to touch a little on this point in your preface.'

Now that the manuscript of the book has been recovered, after being lost for 250 years, we know that Copernicus did not adopt this suggestion. But when Osiander received the manuscript from Rheticus, he had the matter in his own hands, and the recovery of the manuscript has shown what use he made of his opportunity. He added to the title the words 'orbium coelestium' (of the heavenly spheres), and so enveloped the book in a Ptolemaic or even pre-Ptolemaic atmosphere which is lacking in the original title. Next, and much worse, he suppressed Copernicus' original preface, and substituted one of his own writing on the lines of his letter already quoted, so that when the reader opened the book, he was confronted with a preface suggesting that the scheme it described might not be the true scheme of nature, but merely a mathematical fiction which fitted the observations. And for some time after its publication, there seems to have been some doubt as to whether this was all that the new scheme amounted to.

Worst of all, Osiander struck out all mention of Aristarchus; the book as published does not even contain his name, with the result that Copernicus has often been charged with plagiarism, and even with dishonesty, as, for instance, by Melanchthon* and by Erasmus Reinhold.† Nevertheless the original manuscript contained no fewer than four references to Aristarchus, describing him as 'one of those ancient philosophers who, besides the Pythagoreans, regarded the earth as a planet'. We may conjecture that Osiander found

* *Initia Doctrinae physicae*, 1549.
† *Hypotheses Orbium coelestium*, 1551.

the idea of a moving earth so distasteful that he could not endure any mention of the author of the idea.*

After showing that the hypothesis of a moving earth fitted the observations, Copernicus proceeds to answer the objections which Ptolemy had raised against it (p. 94). One was that if the earth rotated once a day, a terrific wind would always be blowing from west to east, so that a bird which once flew out into it would never be able to get back into its nest; Copernicus, of course, explained that the air rotated with the earth. Ptolemy had further objected that so rapid a rotation of the earth would cause everything to fly to pieces; Copernicus pointed out that if the apparent revolutions of the stars about the earth resulted from an actual rotation of the 'sphere of fixed stars', then this sphere would be even more liable than the earth to fly to pieces, because its circumference was greater than that of the earth, and the speed of motion would be correspondingly greater. Finally, ever since the days of the Pythagoreans, it had been argued that the earth could not move in space, since any such motion would result in an apparent motion of the stars—a 'reflected' motion like that seen by the moving child of p. 127. But no such motion was seen; the pattern of the stars stood for ever unaltered, and the constellations did not change their shapes. To this Copernicus answers that the stars are so immensely distant that the relatively minute motion of the earth in its orbit can make no appreciable difference: 'the earth and its orbit stand in the same proportion to the size of the universe as a point does to a clod, or an object of finite dimensions to infinity.' The fact that Copernicus wrote out arguments of this kind shows that he thought, and meant to prove, that the earth was actually rotating and moving through space; his scheme was not a

* See two books in Polish, *Twórca Nowego Nieba* ('The Creator of a New Heaven'), by Jeremi Wasiutynski (1938), and *Nicolas Kopernik*, by Ludvik Birkenmajer (1900). I am indebted to Mr Tadeusz Jarecki for bringing these two books to my notice, and for very kindly translating extracts from them into English for me.

mere 'basis for calculation'. This was also made clear in his dedicatory letter to the Pope, where he speaks of his 'commentaries composed in proof of this motion' of the earth round the sun.

With his refutation of Ptolemy's arguments, Copernicus had proved his case, at least to those few who could rightly assess his arguments. Man could no longer claim that his home was the fixed centre of the universe round which all else revolved; it was one of the smaller of the planets, and like the other planets, it revolved round a far larger sun. If, as man had hitherto believed, he was himself the climax and crown of all creation, then he had been assigned a home in space which was quite incommensurate with his importance, a home, indeed, which stood 'in the same proportion to the universe as a point does to a clod'. Copernicus, it is true, had given pathetically inaccurate estimates of the relative sizes of the earth, the sun and the earth's orbit round the sun, but even so the general principle stood out clear and unassailable —we live on a speck of dust.

Such conclusions might have been expected to make a great stir in the minds of thinking men, but for a time they made none at all. The reason was in part that Copernicus had not only proved his case; he had spoiled it by over-elaboration. Its real strength lay in the majestic simplicity of the central idea—a moving earth replacing a moving sun. Copernicus had so overlaid this central idea with detail that its main advantage seemed to be the reduction of Ptolemy's eighty circles to thirty-four—a reduction in degree but not in kind. The average man could hardly be expected to accept so revolutionary a view of the universe, and one, moreover, which upset some of his most deeply rooted convictions and violated his religious feelings, merely because it changed eighty into thirty-four.

Only a few mathematicians and astronomers expressed their confidence in the new structure of the world, while the majority of men remained hostile or indifferent until the telescope of Galileo began to provide visual confirmation of its accuracy some 66 years later. Even then, one of Galileo's colleagues refused to look through the telescope, on the ground that he saw no reason for reopening a question which had already been settled by Aristotle. He was perhaps exceptional, but many felt genuine objections to the new doctrines on religious grounds. The great astronomer Kepler (p. 164), himself a convinced Copernican, wrote: 'It must be confessed that there are very many who are devoted to Holiness, that dissent from the judgement of Copernicus, fearing to give the Lye to the Holy Ghost speaking in the scriptures, if they should say that the earth moveth and the Sun standeth still.'* Even in 1669, the year in which Newton became Professor at Cambridge, the university entertained Cosimo di Medici with a dissertation against the Copernican astronomy.† And in the eighteenth century, Cassini (1625–1712), the Director of the great Observatory of Paris, and one of the most influential astronomers of his time, expressed himself a convinced anti-Copernican, while the University of Paris taught that the Copernican doctrine was a convenient but false hypothesis. For a considerable period the new American Universities of Yale and Harvard taught the Ptolemaic and Copernican systems on a parallel footing, implying that they were equally tenable. It was not until 1822 that the Roman Church gave formal permission for the Copernican system to be taught as the truth, and not as a mere hypothesis.

TYCHO BRAHE (1546–1601). On 14 December 1546, three years after the death of Copernicus, was born Tycho Brahe, the next great astronomer of this period. In many ways he was the antithesis of Copernicus. The latter had been a great mathematician and a great theorist, but was weak as an observer; Tycho was weak as a mathematician and theorist, but great as an observer—one of the greatest, and perhaps

* *Mathematical Collections and Translations*, Salusbury, 1661.

† Cooper, *Annals of Cambridge*, III, 536.

the greatest, of all time, relative of course to the equipment available in his day.

He was a Dane, the son of a Danish nobleman, although his birthplace, Knudstrup in Scania, now belongs to Sweden. A solar eclipse which occurred on 21 August 1560, while he was a student at the University of Copenhagen, made a great impression on him, and gave him a keen interest in astronomy, so that he began to study the works of Ptolemy and attempted simple observations with crude instruments of his own making. After studying mathematics and astronomy at the Universities of Leipzig, Wittenberg, Rostock and Basle, he made a European tour, and met the Landgrave of Hesse, who was an enthusiastic astronomer. The Landgrave must have been impressed by Tycho's ability, for he persuaded the King of Denmark, Frederick II, to take the young astronomer under his royal patronage. In due course, Frederick granted Tycho a yearly pension and the island of Huen in the straits between Copenhagen and Elsinore, on which to build himself an observatory and a home. Here he built the famous observatory which he called Uraniborg, and furnished so magnificently and equipped so sumptuously that his pension soon proved inadequate, and had to be supplemented by further grants from the King.

When Frederick died in 1588, Tycho's income was reduced, and he left Uraniborg in 1597. Two years later, the German Emperor Rudolph II invited him to Prague, granting him a pension and a castle to use as an observatory. But Tycho's useful life was over, and before he had settled down to serious work, he was struck down by a sudden illness and died on 24 October 1601.

Tycho opposed the doctrines of Copernicus because he thought it contrary both to sound physics and to the clear word of scripture for the massive solid earth to move in space. He was also influenced by the old Ptolemaic objection that the stars did not change their relative positions in the sky, as they would have to do if the earth was in motion.

And so he set to work to improve the Ptolemaic system according to his own ideas. He kept the earth for the centre of his universe, and the Aristotelian sphere of fixed stars for its outer boundary; the sun still circled round the earth, but (and here was the essential novelty) the other planets—Mercury, Venus, Mars, Jupiter and Saturn—all circled round the sun in epicycles. The arrangement is shown in fig. 21; it

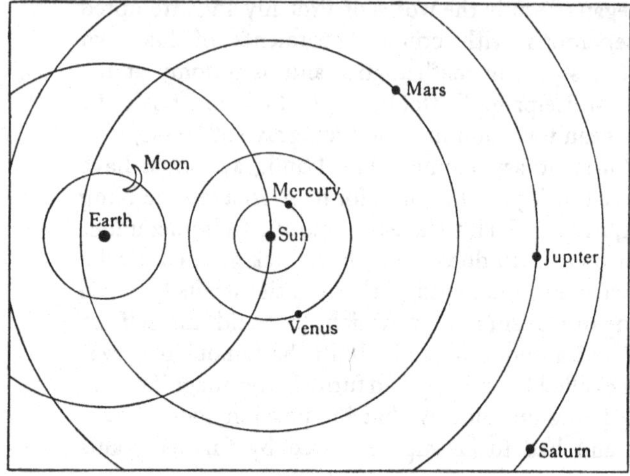

Fig. 21.

gives the same apparent motion for the sun, moon and planets as either the Copernican or the Ptolemaic systems gave in their simplest forms, so that observation could not decide between them. But Tycho's system plays no serious part in the history of science, because all later developments of astronomy were made by believers in the Copernican cosmology.

Tycho's real service to astronomy was as an observer rather than as a theorist; he introduced a new standard of accuracy into astronomy. He attained this in two ways—by the use of better instruments, and by the use of better methods. It may seem an easy matter to obtain greater accuracy by making larger instruments, but actually the situation is not quite so simple. The larger an instrument is, the more it bends under its own weight, and a stage is soon reached when the bending more than neutralises the advantages gained by the increased size of the instrument. Tycho was able to employ large instruments because they were of novel design, being especially planned to escape this objection. He made an even greater advance in the methods of observing. The older astronomers had been content to rely on the best observation they could make; no doubt it would be affected by some small error, but this could not be avoided in an imperfect world. Tycho saw the advantage of taking a great number of observations, all of the same quality, and then averaging the result; accidental errors were now likely to be averaged out.

Using these methods, Tycho determined the more important constants of astronomy with a new accuracy, and made fresh determinations of stellar positions, which were published in the star catalogue of 1602. Probably his observations on the positions of the planets were his best work of all—not for any use he made of them, but for the part they played in later developments. He handed them over to Johann Kepler, an assistant whom he had engaged just before his death, with results that we shall see later. But Tycho was an improver rather than an originator; he plays a great part in the history of astronomical technique, but figures little in the history of thought.

Nevertheless, some of his work reached out beyond the merely technical problems of astronomy. On the evening of 11 November 1572, he observed a bright new object in the constellation of Cassiopeia. We know now that it must have been a 'nova' or new star; these objects appear at fairly frequent intervals, flashing out very suddenly and gradually fading back into obscurity. But to Tycho it was 'a miracle indeed, either the greatest of all that occurred in the whole range of nature since the beginning of the world, or one certainly that is to be classed with those attested by the Holy

Oracles, the staying of the Sun in its course in answer to the prayers of Joshua, and the darkening of the sun's face at the time of the Crucifixion'.*

If this new object had belonged to the solar system, it would have appeared to move against the background of the fixed stars, as the planets do. As Tycho could observe no such motion, he concluded that the object must belong to the 'sphere of fixed stars'—in brief, it must be a star. The Aristotelians had taught that everything in these outer regions of space was perfect, and therefore unchanging: 'All philosophers agree, and facts clearly prove it to be the case, that in the ethereal region of the celestial world no change, in the way either of generation or corruption, takes place; but that the heavens and the celestial bodies in the heavens are without increase or diminution, and that they undergo no alteration.'† Tycho, by showing from direct observation that these regions were no more immune from change than the regions nearer to the earth, had dealt a shattering blow to the Aristotelian cosmology.

GIORDANO BRUNO (1547–1600). The prominent scientists of the last few centuries had been mainly ecclesiastics, and most of them had occupied high positions; perhaps this is not surprising, since knowledge and learning belonged almost exclusively to the Church. The scientist to whom we come now was of a very different type.

Giordano Bruno had been born at Nola near Naples in 1547, and became a Dominican at the age of 15. He was a man of independent mind, of an aggressive, intolerant and turbulent spirit, with more than a touch of the mountebank and charlatan in his make-up, so that we can imagine that he was a cause of some trouble to his monastic superiors. Learning that he was under suspicion for heresy on the subjects of transubstantiation and the immaculate conception, he fled from Italy to wander in France, England, Germany and Switzerland. After teaching in the Universities of Lyons,

* 'De nova Stella', *Opera*, vol. I. † Ibid.

Toulouse, Montpellier and Paris in turn, he finally came to London in 1583. Here he published three small books in Italian, bearing false Venetian imprints, of which one, *Dell' infinito Universo e Mondi* ('On the infinite Universe and its Worlds'), is of special scientific interest.

In philosophy, Bruno was something of a pantheist. He saw nature as a world of life and beauty, full of activity and pulsating with divinity. And, since there is nothing finite about God, so there can be nothing finite about the universe. He wrote: 'It has seemed to me unworthy of the divine goodness and power to create a finite world, when able to produce beside it another and others without end, so that I have declared that there are endless particular worlds similar to this earth; with the Pythagoreans I regard it as a star, and similar to it are the moon, the planets, and other stars, which are infinite in number, and all these bodies are worlds.' In another place he explains that each world has its own sun round which it moves.

In this way, Bruno carried astronomy out beyond the solar system, and introduced the modern view of the system of the stars. He was treading the road which had been opened by Nicolas of Cusa and Copernicus, but he was incomparably more revolutionary than either. He displaced not only the earth, but also the sun, from the centre of the universe—in fact, there no longer was any centre, for 'as the universe is infinite, no body can properly be said to be in the centre of the universe or at the frontier thereof'. Man's home in space occupied no preferential position, and could expect no preferential treatment; all the planets circling round all the suns stood on the same footing; all were evidence of the goodness of God, and sometimes he thought that all were God.

The Church had passed over the revolutionary doctrines of Copernicus without showing any active disapproval, but this new revolution touched its interests much more closely. Religion meant nothing unless the Creator was distinct from His creation; Bruno was preaching that they were identical.

It was essential for the Church to have room for a heaven and a hell; it had so far placed hell inside the earth, and heaven beyond the 'sphere of the stars'. Bruno's new cosmos left no room for a material heaven. Copernicus' doctrines had not called for a restatement of any of the fundamental doctrines of religion; the new doctrines of Bruno called for a restatement of many, unless God was to become a mere tribal god of the Planet Earth. Living though he was on a moving planet, man might still have been the centre of God's interest, the main concern of his Creator; Bruno's doctrines now implied that there were infinite other worlds of the same kind which might share the interest of the Creator. All this was too antagonistic to the established doctrines of the Church to be passed over in silence.

In 1593 Bruno was imprudent enough to return to Italy, and the Inquisition were informed where he was. They captured him, kept him in prison for seven years, and finally tried him on a number of counts. Finally, the judgement was pronounced: he was to be 'punished with all possible clemency, and without shedding of blood', a formula which in practice meant death by burning at the stake. Bruno is said to have commented to his judges: 'Perhaps you who condemn me are in greater fear than I who am condemned.' Some have thought this trial and judgement one of the most shameful incidents in the record of the Church; others remind us that we do not quite know on what grounds Bruno was condemned, since it was not the practice of the Inquisition to make public the grounds on which a judgement was based. He was certainly charged with more than unorthodox scientific opinions; he had denied the doctrines of transubstantiation and of the immaculate conception, and had written a pamphlet, *On the Triumph of the Beast*, in which the title-part was assigned to the Pope.

So Bruno died, and henceforth he could influence human thought only through the meagre writings he had left behind him. The chances seemed small, but the improbable came to pass. In the very year of his death another book appeared in which the same views were advanced, no longer by an obscure monk but by a writer of authority and position—William Gilbert (1540, or perhaps 1544 or 1546, to 1603), the personal physician to Queen Elizabeth. The book, *De Magnete*, dealt mainly with physics, and has indeed become famous as one of the corner-stones on which modern electrical science has been built. But its last chapter describes a hypothetical scheme of the universe; it is the scheme of Bruno, although his name is nowhere mentioned. This might be explained as two minds thinking alike were it not that in 1651 a posthumous book by the same author appeared,* in which the same ideas are advanced as in the earlier book, and are now definitely attributed to Bruno. In this and similar ways the spirit of Bruno lived on, and in its own time produced even greater changes in thought than the hypotheses of Copernicus.

The year 1600, the closing year of the century, the year of Bruno's death, the year of publication of *De Magnete*, the year in which electrical science was born, forms a fitting milestone at which to close the present chapter, but we must first record the developments which had taken place up to this time in other sciences than astronomy.

MECHANICS

The sixteenth century also produced noteworthy advances in mechanics, which had progressed but little since the time of Archimedes. This was now established on a firm basis, largely through the investigations of two men—the Fleming Stevinus of Bruges, and the Italian Galileo Galilei. Although they were almost contemporaries, the two men worked independently, and their results supplemented one another to make a solid foundation for a new science of mechanics. Stevinus was concerned mainly with the mechanics of objects at rest (statics), and Galileo mainly with the mechanics of objects in motion (dynamics).

* *On our Sublunary World, a New Philosophy.*

STATICS

STEVINUS (1548–1620) was an engineer who had attained to high rank in the Dutch army. His most striking achievement was the discovery of the law which we now call the 'Parallelogram of Forces'.

It is rare for any object to be under the action of only one force; more usually many forces are at work at the same time. A falling leaf, for instance, is acted on by the gravitational attraction of the earth (the 'weight' of the leaf), by the resistance of the air, and probably also by the force of the wind. If

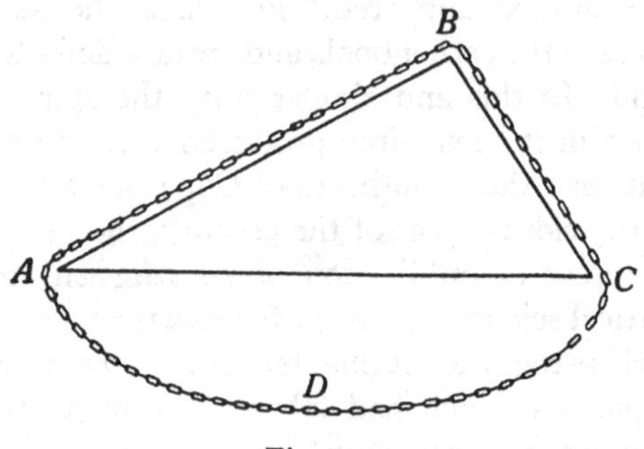

Fig. 22.

it were acted on by its weight alone, it would fall vertically to the ground, but if the pressure of an east wind is added, it will be blown westward, and will fall farther to the west than it would otherwise have done. The question is: how much farther? Or in more general terms, how are we to estimate the combined effect of two or more forces when they act together?

Stevinus did not experiment to find the answer, but considered an ideal experiment of which the result could easily be foreseen. He imagined a wedge like *ABC* in fig. 22 to be firmly fixed with the longest side *AC* horizontal, and to have an endless uniform chain *ABCD* strung round it. Without experimenting, Stevinus knew that the chain would lie at rest as in the diagram. For the only conceivable alternative was perpetual motion, and it had been a commonplace in science from the time of the Greeks down to Leonardo that this could not occur.

Stevinus next assumed—this time from intuition, and not from either reasoning or experiment—that the hanging part of the chain, *ADC*, could be cut away without disturbing the equilibrium of the remainder. If this were done, pieces *AB*, *BC* of the chain would stand in equilibrium. As the weights of these pieces were in proportion to their lengths, Stevinus inferred that any two objects standing on the faces *AB* and

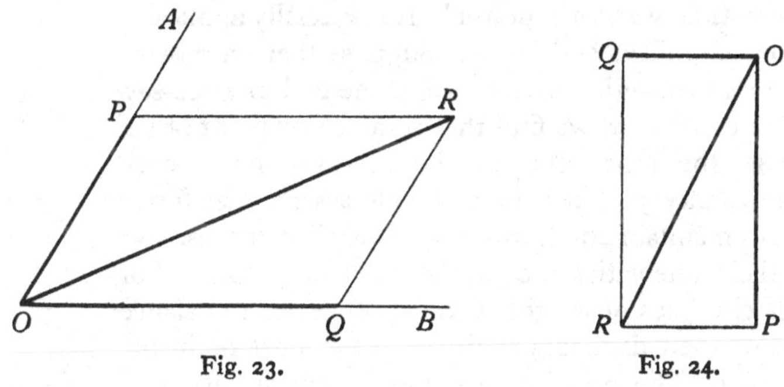

Fig. 23. Fig. 24.

BC and connected by a string would be in equilibrium if their weights were in the ratio of the lengths *AB* and *BC*. From this, simple mathematics led to a rule for determining the effect of two forces acting simultaneously on the same object. The rule is as follows.

Suppose that two forces act simultaneously on an object at a point *O*, and that they act in the directions *OA*, *OB* in fig. 23. From the lines *OA*, *OB* we cut off lengths *OP*, *OQ* proportional to the intensities of the two forces, and complete the parallelogram *OPQR*. Then the rule tells us that the two forces will have the same effect as a single force of intensity proportional to the length *OR*, acting in the direction of *OR*.

If, for instance, *OP* in fig. 24 represents the weight of a falling leaf, and *OQ*, on the same scale, represents the force of the wind on the leaf, then the leaf will fall as though it were acted on by a single force proportional to *OR*.

This ingenious argument rested on a mixture of experimental knowledge (as to the impossibility of perpetual motion), of intuition, and of assumption. It was important in two ways: it clarified the idea of a body being under the action of a number of forces simultaneously, and it yielded a result which was indispensable for the further progress of mechanical science.

Stevinus also enunciated the principle of 'virtual displacements', which is popularly expressed in the saying that 'What we gain in motion, we lose in power'. It is specially applicable to problems of pulleys and levers. Suppose that on moving one point of a mechanical system—e.g. one end of a see-saw—through a distance x, we find that some other point of the system—e.g. the other end of the see-saw—has moved through a distance y. Then the principle asserts that forces X and Y can maintain equilibrium when applied at these two points if their intensities are in the ratio of y to x. For example, if two boys of weights 6 and 7 stone are to balance on a see-saw, their distances from the pivot must be in the ratio of 7 to 6. The principle was not a new discovery of Stevinus, for it had been vaguely known to Aristotle and Archimedes, and in a different form to Leonardo.

Dynamics

Stevinus explains all this in a book *Statics and Hydrostatics* which he published in 1586, and further describes how he and Grotius had experimented on the fall of objects under gravity, and found that when a light weight and a heavy weight were dropped from the same height they took the same time to reach the ground. This was contrary to the Aristotelian physics, which had taught that objects were light or heavy by their intrinsic nature, and that all substances had their natural places in the universe which they strove to reach with varying degrees of success, so that heavy bodies fell and light bodies rose with speeds depending on their lightness and heaviness. Aristotle would have asserted that bodies of different substances would not reach the ground at the same time; Stevinus and Grotius now proved that they did.

GALILEO (1564–1642). This was all very important, but the contribution of Galileo was even more so. Galileo Galilei was born at Pisa on 18 February 1564, the day on which Michael Angelo died, his father being an impoverished nobleman, whose main occupations and interests were music and mathematics. He was educated at the Monastery of Vallombrosa, where we may be sure that he received a conventional Aristotelian education; he studied Greek, Latin and logic, but showed a distaste for science. Then his seniors urged him to become a novitiate of the order, and his father, demurring, sent him to the University of Pisa to study medicine. A lecture on geometry, which he heard by accident, convinced him that mathematics was far more interesting than medicine; he is reported to have hung about the door of the mathematical classroom to pick up such crumbs as he could of the knowledge which was being imparted within. When the authorities heard of this, they transferred him from medicine to mathematics and science. In 1585 he had to leave the university from lack of funds, and returned to Florence, where he lectured and made such a scientific reputation that he was appointed to a lectureship in his old University of Pisa at the early age of 25. But an independent mind, a sarcastic temperament and a sharp tongue soon made him unpopular with all whose opinions differed from his own. He held his lectureship for only two years, and was then appointed to the Professorship of Mathematics at Padua, a position which he held for the next 18 years.

At Pisa he set to work to discover the true principles of mechanics, having long felt convinced that the conventional Aristotelian doctrines on the subject were erroneous. He began by dropping bodies of different weights from the same height—according to one account, from the top of the leaning tower at Pisa—to test the Aristotelian doctrine that different bodies fell at different rates, and found that a cannon-ball and a musket-ball took the same time to fall, a confirmation of the result which Stevinus and Grotius had obtained at Delft. Here was conclusive proof that something was wrong with the Aristotelian mechanics; Galileo set to work to find what it was.

The speed of a falling object obviously increased throughout the fall, as Leonardo had insisted, and Galileo first tried to find the law which governed this increase. His first conjecture was that the speed at each point might be proportional to the distance through which the body had fallen, but he soon found that if this were the true law of fall, a body would never get started; it could literally stand suspended in the air without falling at all. He next conjectured that the speed at each instant might be proportional to the time that had elapsed since the body had been set free, and set to work to test this conjecture. It was, of course, hopeless to try to measure either the speed or the time of fall directly. But Galileo saw that if his conjecture were sound, then the speed at any point would be just double the average speed up to that point, and this average speed could be obtained by dividing the distance of fall by the time of fall. In theory, then, Galileo could test his conjecture by measuring the times taken by an object to fall through different heights—yet how was he to measure such short periods of time, when the only methods known for measuring time were by sundial, by burning candles or oil-lamps, by sand-glasses and water-clocks, and by very crude mechanical clocks?

Galileo improved the water-clock in a very ingenious way, by letting the water drip into a receiver and then weighing the amount which had fallen with great accuracy, but the times to be measured were still uncomfortably short. Galileo accordingly slowed his experiments down, by substituting a slow roll down a gentle slope for the rapid vertical fall, in the belief that the same laws must govern both, as indeed is the case. He set up a gently sloping plank, some 12 yards in length, and made polished steel balls roll down a narrow groove cut into it. With this simple apparatus he was able to verify his conjecture that the speed of fall increased uniformly with the time—the law of 'uniform acceleration'. It was one of the great moments of the history of science.

For it now became clear that the effect of force was not to *produce* motion, but to *change* motion—to produce acceleration, and a body on which no force acted would move at a uniform speed.

The Aristotelians had taught that all motion needed a force to maintain it, so that a body on which no force acted must needs stand at rest. In accordance with these ideas, Aristotle had himself introduced his Unmoved Mover (p. 65), God Himself, to keep the planets in motion, while the medieval theologians had postulated relays of angels for the same purpose. It now appeared that to keep a body in motion it was only necessary to leave it to itself; a body acted on by no force would not in general stand at rest, but would move with uniform speed in a straight line, because there would be nothing to change its motion. Galileo checked this by letting his rolling steel balls continue their motion on a horizontal plane; they moved with undiminished speed until they were checked by friction and the resistance of the air.

This last observation was not altogether new; it was too simple and obvious for that. Others had noticed that a rolling ball would continue its motion for some time, but had adduced this as a proof of the naturalness of circular motion—a rolling sphere persisted in its motion because every particle of it was then moving in a circle, but if the rolling object was irregular in shape, so that the circular motion was impossible, then the motion soon stopped.

Neither was the idea new that motion tended to persist in the absence of all force. We have seen that Leonardo had declared that 'every body has weight in the direction in which it is moving', while Plutarch had put the thing even more clearly when he wrote (A.D. 100) that 'everything is carried along by the motion natural to it, if it is not deflected by something else'.* But Galileo was the first to establish the principle experimentally; where others had conjectured, Galileo proved.

Yet, strangely enough, he never announced the principle with perfect clearness. Perhaps Descartes was the first to do so, when he wrote (1644): 'When a body is at rest, it has the power of remaining at rest and of resisting everything which could make it change. Similarly when it is in motion, it has the power of continuing in motion with the same velocity and in the same direction.' Thirty years later, Huygens restated it in the form: 'If gravity did not exist, nor the atmosphere obstruct the motion of bodies, a body would maintain for ever, with equable motion in a straight line, the motion once impressed on it.'† In 1687 Newton again restated it in his *Principia* (p. 190), and made it the foundation of his whole system of dynamics. But the main credit for the law which was to revolutionise the whole of mechanical science must go to Galileo and his experiments.

Galileo next discussed how a body would move when forces acted on it in some direction other than that of its motion; a projectile moving through the air while gravity acts on it is an obvious instance. Galileo proved that if the resistance of the air were negligible, then the path of every projectile would be a parabola, one of the conic sections on which the Greeks had expended so much patient labour; these curves now re-entered science as essential parts of the great scheme of nature, and not as mere abstractions of the mathematicians.

In this case the resistance of the air is not negligible, and Galileo did not know how to allow for it. But in the next problem which he studied, air resistance was too slight to introduce any complications. In his early days at Pisa, he had

* *De Facie in Orbe Lunae.*

† *Horologium Oscillatorium*, 1673.

Galileo's design of a pendulum clock

From Favaro's 'Galileo e Cristiano Huygens. Nuovi documenti sull' applicazione del pendolo all' orologio' (*Nuovi studi Galileiani*, Venice, 1891)

watched the roof-lamp of the Cathedral swinging in the breeze, and noticed that a small swing took as long a time as a great one. Here he had had no clock except his own heart, and had timed the oscillations by counting his pulse. He now confirmed these primitive observations by exact experiment in the laboratory. Not only did the time of swing of a pendulum prove to be the same whether the swings were great or small, it was also the same, whatever the material of which the swinging mass was made; other things being equal, a ball of lead would swing to and fro in the same time as a ball of cork. This showed that gravity caused all substances to increase their speed of fall at the same rate.

Late in his life, Galileo saw that this property of a swinging pendulum might make it possible to construct a clock to keep better time than the crude instruments then in use. The main difficulty was to devise some means of maintaining the motion of the pendulum, possibly by drawing on some external source of power, as, for instance, a falling weight. Galileo thought that he had solved this problem, but he never constructed such a clock himself, and neither did his son Vincenzio nor his pupil Viviani to whom he gave instructions; it was first done by Huygens who patented his device in 1657, and described it in his *Horologium Oscillatorium* (1673).

Only two of the five parts of this famous book dealt directly with clocks; the rest contained much that was new in general mechanics. Most important of all, it contained a discussion of the so-called 'centrifugal force'—the force which a whirling sling exerts on the string by which we hold it. Huygens showed that centrifugal force in general is proportional to the square of the speed of the moving body and inversely proportional to the diameter of the circle in which the body moves. Its amount per unit mass is given by the well-known formula $F = v^2/r$, which Newton used to calculate the gravitational force which the sun must exert on a planet to counteract centrifugal force and keep the planet moving in its circular orbit.

In this and similar ways the mechanics of solid bodies was put on a sound theoretical and experimental basis, the statical part mainly by Stevinus, and the dynamical part by Galileo.

Hydrostatics

The same story was repeated in the mechanics of fluids, which, when Stevinus and Galileo first appeared on the scene, still stood much where Archimedes had left it.

The Aristotelians had said that the shape of a body determined whether it would sink or float in water; a needle, for instance, or a leaf floated, while a cube or a sphere sank. Archimedes had known better than this; his famous experiment with the crown had depended on the density or 'specific gravity' of the metal, and Archimedes understood that it was this, rather than its shape, that determined whether an object would float or sink. The Arabians had been familiar with this idea, and had determined the specific gravities of a number of substances.

Galileo now performed a magnificently simple experiment which settled the question once and for all. He let a ball of wax sink to the bottom of the water in a vessel, and then increased the density of the liquid by adding salt to the water. When the density reached a certain value, the ball of wax was seen to float up to the surface of the liquid. A body, then, did not sink or float according to its shape, but according to its density relative to the fluid in which it was immersed.

Stevinus now studied the conditions inside a mass of fluid. Fluid substances may be divided, broadly speaking, into viscous (or sticky) substances like pitch or treacle, and non-viscous liquids like water or wine. Fluids of the latter kind have no cohesive force, whence it can be shown that, if the liquid is standing at rest, there must be a perfectly definite pressure at every point, this being the force exerted by the liquid on a unit area of surface, independently of the orientation of the area. This idea of a definite pressure at every point had been introduced into science by Archimedes and revived by Leonardo. Stevinus now showed that the pressure at any point in a perfectly non-viscous liquid would depend only on the 'head' of liquid above the point, so that it would, for instance, be the same at any two points both of which were 10 ft. below the surface of the sea. This law provided a simple explanation of the so-called 'hydrostatic paradox'—that the thrust which a liquid exerts on the floor of a containing vessel depends only on the area of the floor and its depth below the surface of the liquid, but does not depend on the shape of the

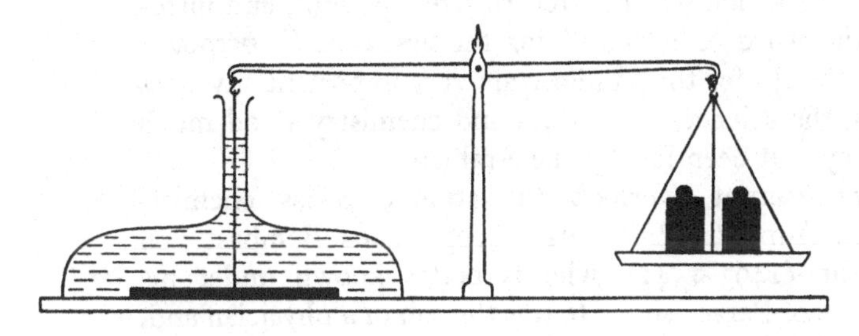

Fig. 25.

vessel. Stevinus proved these laws by laying a close-fitting plate on the floor of his containing vessel, and measuring what pull was required to raise this against the pressure of the water in the vessel. By experimenting with vessels of different shapes, Stevinus verified the result stated above. By using a vessel which terminated in a long narrow tube, as in fig. 25, a slight weight of water could be made to exert a very great pressure; Stevinus remarked that 1 lb. of water could be made in this way to exert a pressure equal to the weight of 100,000 lb. of water. This is the principle used in the 'hydraulic ram' and the hydraulic brakes of motor-cars.

In this way the science of hydraulics was put on a sound basis of theory and experiment, although its main developments did not occur until the next century.

PHYSICS AND CHEMISTRY

There is little enough to record in the other sciences. In physics the outstanding event was the publication of Gilbert's *De Magnete* (p. 141) in 1600. Its full title was *De Magnete magneticisque Corporibus et de magno Magnete Tellure* ('On the Magnet and Magnetic Bodies, and on that great Magnet the Earth'), but it dealt with more subjects than are enumerated in this title, providing a foundation for electrical science, describing various phenomena and experiments in statical electricity to which we shall return later (p. 276), and introducing the word 'electricity' for the first time (ἤλεκτρον = amber). Except for this, and certain very important advances in optics, the sciences of physics and chemistry stood much where they had been left by the Arabians.

Yet we cannot overlook the strange Swiss chemist-physician, Aureolus Philippus Theophrastus Bombast von Hohenheim (1493–1541), who is better known under his Latin name of Paracelsus. He was the son of a physician and, after studying medicine at Basle and Wurzburg, was himself appointed Professor of Medicine at Basle in 1526. Personally he was insufferably conceited, arrogant and boastful; on being appointed Professor, his first act was to show his contempt for his great predecessors Galen and Avicenna by publicly burning their works. His own works, incidentally, are badly written and incredibly difficult to understand. Yet he may almost claim to have been the first true chemist in the whole history of science.

His interest was mainly in the curative uses of chemistry or, as would then have been said, in iatro-chemistry. We have seen how the Arabians, following a lead given by Jabir-Ibn-Hayyan (p. 105), had replaced the four Pythagorean elements by three 'principles', which they described as sulphur, mercury and salt. Sulphur did not mean our chemical element of atomic weight 32, but that quality of a substance which made it combustible, or, as we might now say, its hunger for oxygen. Salt meant that quality which enabled a substance to resist the action of fire, or the residue which remained after calcination, while mercury was used to signify the characteristic quality of metals. Gold, for instance, was said to contain a very pure mercury; copper contained no sulphur, but much salt and mercury, and so on.

Paracelsus accepted most of this and also believed in the 'vital spirits' of early Greek thought. He was a firm believer in astrology and was convinced that different parts of the body were associated with the planets; for example, the liver with Jupiter and the brain with the Moon. Digestion, he said, was caused by an independent spirit, called Archaeus, in the stomach. He then proceeded to develop a Chemo-therapy of his own, asserting that every organ of the body had its own brand of vital spirits, and that maladjustment of these led to all the various ailments which affected the human body, so that these could be cured by dosing the sufferer with the right chemical substance.

This led to his trying the effects of various chemicals, both poisonous and non-poisonous, on the bodies of his unfortunate patients, and this in turn led to his expulsion from Basle. But before this happened, he learned how to prepare a great number of hitherto unknown chemicals, and this gave a start to modern chemistry, which began to disentangle itself from the alchemy which had hitherto usurped its place.

Paracelsus was the first person, apparently, to use the name 'alcohol' for strong spirits of wine. Air he called 'chaos', probably for astrological reasons rather than because he realised its composite nature. He seems to have prepared ether, to which he gave the generic name 'sulphur' (prepared as an 'extract of vitriol'), and discovered its anaesthetic properties, without realising that he had made one of the most useful discoveries of medical science. Chickens, he found, could be put to sleep with it, and awakened uninjured 'after a moderately long time'. A few years later another physician, Valerius Cordus (1515–44), again described how to prepare ether from sulphuric acid and alcohol, but still its practical value as an anaesthetic passed unnoticed.

This period saw the invention of two of the most useful of scientific instruments, the microscope and the thermometer. The invention of the microscope is generally attributed to Zacharias Jansen, a spectacle-maker of Middelburg in Holland, who is said to have discovered the essential principle accidentally. But his instruments were more like small telescopes than what we now call a microscope, a double convex lens acting as an object glass and a double concave lens as eyepiece. The combination was mounted in a tube of perhaps 18 in. length and 2 in. diameter.

The thermometer was invented by Galileo, according to his disciple Viviani, in or about the year 1592. We ought perhaps to say reinvented, for Hero of Alexandria had understood the principle, and had used it for some of his mechanical toys, while a certain Philo of Byzantium is said to have made similar instruments at about the beginning of the Christian era. Galileo's thermometer consisted of a glass bulb, 'about the size of a hen's egg', from one point of which a thin tube was drawn out, no thicker than a straw, but several inches long and open at the far end. To use the thermometer the open end of the tube was placed in a vessel of water. If the bulb was now heated, the air in it expanded so that some of it forced its way out of the tube and through the water. As the bulb cooled again, water was sucked up into the tube, and its amount showed by how much the bulb had been heated. The instrument did not, of course, measure the absolute temperature of the hot bulb, but only the difference between this and the temperature from which the bulb started. It was also sensitive to changes in atmospheric pressure. Within a few years of its invention, the instrument was in use as a clinical thermometer, the patient being told to put the egg-sized bulb in his mouth.

MATHEMATICS

More important in many respects than the developments in physics were the developments in mathematics. It was an age of useful, rather than of great, mathematicians, and the new knowledge gained was perhaps less important than the new methods of acquiring knowledge.

Prominent among the mathematicians of the period was Niccolo Fontana (1506–59), who was better known as Tartaglia (the stammerer). As a boy his skull had been broken and his jaws and palate cut open when the French had sacked his native town of Brescia—hence the defect of speech from which he got his nickname. After becoming Professor of Mathematics at Venice, he published a book (*Nova Scientia*, 1537) in which he investigated the motion of bodies under gravity and discussed the range of projectiles, stating that the range is greatest when the projectile is fired at an elevation of 45°, which would be accurate if air resistance could be neglected. He also published a treatise on numbers,* in which he showed how to derive the expansion of $(1+x)^{n+1}$ from that of $(1+x)^n$, and so took the first step towards the binomial theorem.

In 1530 he stated that he could solve a restricted type of cubic equation, i.e. an equation involving x^3, as well as the x^2 and x of the ordinary quadratic equation which Diophantus had solved—and a certain Antonio del Fiore challenged him to a mathematical contest in which each contestant was to set the other thirty problems involving cubic equations, and the stakes were to go to whoever solved most. On receiving this challenge, Tartaglia set to work and discovered the general solution of all cubic equations, so that he won with the greatest ease.

But now another mathematician appeared on the scene—Girolamo Cardan (1501–76), a man of brilliant ability but unstable mind. He was a physician by profession, holding

* *Trattato generale di numerì et misure* (1556).

Chairs of medicine in the Universities of Milan, Padua and Bologna in turn, but made himself a European reputation by his works on astrology and algebra. His principal work *De Varietate Rerum* (1557) is noteworthy for its suggestions for teaching the blind to read and write by their sense of touch, and for teaching the deaf to converse by the use of sign-language.

When the result of the contest between Tartaglia and Antonio del Fiore was made known, this man Cardan implored Tartaglia to show him how to solve cubic equations, and obtained the secret, after giving a pledge that he would keep it to himself. Some fifteen years later he published it in his treatise on algebra,* but without any mention of Tartaglia, so that to this day the solution is commonly described as Cardan's solution. When Tartaglia made a stir, Cardan excused himself on the ground that he had only been given the result and not the method of solution. Tartaglia, nevertheless, challenged him to a duel, the weapons to be—mathematical problems. Cardan failed to appear at the appointed time, and the meeting ended in disorder.†

The biquadratic equation (which involves x^4 as well as x^3, x^2 and x) was first solved by Ferrari, a pupil of Cardan, in 1540, and Cardan published it in the same book. Equations of still higher degree do not admit of exact solution, as various mathematicians proved in the nineteenth century.

Cardan's book had a more serious interest, since it contained the first known discussion of 'imaginary' quantities, i.e. quantities involving the square root of -1. The layman in mathematics may feel but little interest in imaginary

* *Artis magnae, sive de Regulis Algebrae, liber unus.*

† The solution of the cubic is a simple matter to have caused such a commotion. Any cubic equation $x^3+px^2+\ldots=0$, can be reduced to the form $y^3+qy=r$ by substituting $y-\frac{1}{3}p$ for x. On comparing this equation with the simple formula $(a-b)^3+3ab\,(a-b)=a^3-b^3$, we see that $a-b$ will be a solution of the equation if a and b are chosen so that $3ab=q$ and $a^3-b^3=r$. Writing the first of these equations in the form $a^3b^3=\frac{1}{27}q^3$, we readily find the values of a^3 and b^3, and hence of $a-b$.

IV

Tycho Brahe's mural quadrant

quantities on the ground that they cannot exist, or at least cannot be understood; he probably means that they cannot be represented pictorially, like real quantities such as 2 or −2. But they play just as important a part in modern mathematics as do real quantities, and they are of special consequence to the physicist because of their occurrence in the theories of wave-motion, alternating currents of electricity, and relativity. The introduction of imaginary quantities into mathematics was the beginning of a great extension of the subject, the end of which cannot yet be seen. Cardan shows amongst other things that an equation may have either real or imaginary roots, and that the imaginary roots, if any, always occur in pairs.

Shortly afterwards, an Italian mathematician Bombelli treated the same subject in rather more detail, but after him the subject of imaginary quantities had to wait two centuries, when it received exhaustive treatment at the hands of Euler (p. 231), Gauss (p. 279) and others. Bombelli did further useful work in improving algebraic notation, although his efforts were less successful than those of Vieta, to whom we now turn.

FRANCOIS VIÈTE (1540–1603), better known under his Latinised name of Vieta, was a French lawyer and public servant who devoted his leisure to mathematics, in which he ultimately achieved a considerable reputation. This rested in part on his success in solving a problem involving the solution of an equation of the 45th degree which Adrian Romanus, Professor of Mathematics in the University of Louvain, had issued as a challenge to the world. Like many of these 'challenge' problems, it was showy rather than solid. Romanus knew the ordinary formula for $\sin(A+B)$ (p. 93), and from this it was easy to calculate $\sin 2A$, $\sin 3A$, $\sin 4A$, etc., in succession. Actually Rheticus had given the values of $\sin 2A$ and $\sin 3A$, but Romanus had apparently calculated values up to $\sin 45A$, expressing them in powers of $\sin A$, and the equation of his problem was simply $\sin 45A = 0$, written in the disguise of a power series.

King Henry IV of France brought the challenge to the notice of Vieta, who by a stroke of luck had also worked out the general formula for sin nA, and so was able to hand the King a solution within a few minutes; it was of course sin 4°. This success and the decoding of a despatch in cypher gained for Vieta a great reputation in France, but our interest in him is that he was largely responsible for our present algebraic notation, and for the introduction of decimals into calculations.

Up to the time of Bombelli, some mathematicians had used entirely distinct symbols, as, for instance, A, B, C, D, ..., to denote the quantities we now write x, x^2, x^3, x^4, ..., while others had used R or Rj (radix), Z or C (census), C or K (cubus). Then Bombelli introduced the symbols $\underset{\smile}{1}$, $\underset{\smile}{2}$, $\underset{\smile}{3}$, $\underset{\smile}{4}$. Vieta now replaced these by A, A quadratus, A cubus, A biquadratus, etc., and subsequently improved this to A, Aq, Ac, Aqq, etc., so giving to algebra pretty much its present form. His introduction of our present decimal notation for fractions was perhaps even more useful. The Flemish mathematician Simon Stevinus (1548–1620) had already introduced such a notation, but it was awkward and unwieldy; the number which we write as 3·1416 was written first as

3⓪ 1① 4② 1③ 6④

and later as 3, $1'4''1'''6^{\text{iv}}$. Vieta introduced the far simpler notation 3,1416, which is still used on the continent, although we have replaced the comma by a stop. This last change, so far as we know, was introduced in 1616 by Napier's friend Henry Briggs (1561–1630), who was then Gresham Professor of Astronomy in London, and subsequently became Savilian Professor at Oxford.

A far greater achievement was the invention of logarithms by another amateur, John Napier (1550–1617) of Murchiston, Scotland. He was a man of wealth and position who had chosen religious and political controversy for his main occupation in life, and mathematics for his hobby. Vieta's use of powers might well have drawn renewed attention to the formula $A^m \times A^n = A^{m+n}$, which was as old as Archimedes (p. 76), and from this it would have been but a small step to the central conception of logarithms. But Napier does not seem to have had any flashes of insight of this kind, and it was only after many years of thought and labour that the idea of logarithms dawned upon him. He compared numbers and their logarithms with the sequences of terms in geometrical and arithmetical progressions. He sent a preliminary statement of his discovery to Tycho Brahe in 1594, but did not make it public until 1614, when he described it in a book *Mirifici Logarithmorum Canonis Descriptio*. This contained laboriously calculated tables of what we should now call log sines and log tangents. Napier's ambition had been limited to shortening trigonometrical calculations, and he had not concerned himself with the operations of ordinary arithmetic. The possibility of facilitating these was first realised by Briggs. Thus while we must credit Napier with the discovery of the general principle of logarithms, the credit for making them into the everyday working tool of the mathematician must go largely to Briggs. Between them, the two men gave a great gift to the world of science.

CHAPTER VI

THE CENTURY OF GENIUS

(1601–1700)

HERE and there, in the history of human thought and action, we find periods to which the epithet 'great' may properly be applied—in Greece the fourth century before Christ; in England the Elizabethan age; in the domain of science the seventeenth century, the 'century of genius', to which we now come.

It would be very undiscerning to suppose that such a period of greatness could arrive as a mere accident, a specially brilliant galaxy of exceptional minds just happening to be born at one particular epoch. Mental ability is believed to be transmitted in accordance with the laws of heredity, in which case the laws of probability will see to it that no abrupt jump occurs from one generation to the next. Thus a period of greatness must be attributed to environment rather than to accident; if an age shows one particular form of greatness, external conditions must have encouraged that form. For instance, the sixteenth century was an age of great explorers because conditions then specially favoured exploration; the pioneering voyages of Columbus, Vasco da Gama, Cabot, Magellan and others had drawn attention to the wealth of new territory awaiting discovery, while men had learned to build ships which could defy the worst fury of the ocean.

Perhaps the reasons which made the seventeenth century a great scientific age were somewhat similar—a realisation that vast virgin territories were awaiting exploration and development, especially in the physical sciences where direct experiment and observation were replacing a rapidly decaying faith in authority, and that the requisite tools were becoming available as they were needed; for it had become evident that the unaided human senses were inadequate to explore the deeper secrets of nature. Thanks to the Arabian physicists, to Roger Bacon and others, the general principles of optics were well understood; at the beginning of the century the microscope was already in existence, the telescope was soon to come, and other instruments followed in rapid succession. In mathematics logarithms had just been discovered, with their power to replace a lifetime of labour by a few hours of work.

The Church had almost withdrawn its age-long opposition to the study of science. From the time of Anaxagoras on, religion had been at the best unsympathetic, and more often openly hostile, to science; in the Middle Ages it had been the main brake on progress. Thought was then dominated by religion to an extent which it is hard to imagine to-day. The universe was regarded as a wheel of many spokes, but all radiated outward from man and his earth, and all led in the minds of most men to God and His heaven—or hell. Then came the renaissance, dislodging men's thoughts from their accustomed groove, giving them a wider vision, including one of a world in which Christianity had not even existed. Men saw that the external world was worthy of study—to some for its own sake, to others as evidence of the beneficence of its Creator. The intense preoccupation with the details of religion began to pass away, and science became free to find the path to the truth by its own methods.

One indication of the new and more favourable position of science was the foundation of scientific academies, many of them national in their scope and enjoying royal patronage. In the academies of antiquity learned men had been able to discuss their problems with one another and with their students. The medieval universities had been but poor substitutes for these, often being too much under the control of the Church for science to be viewed with much favour. When the sixteenth century brought its general revolt against authority, the need was felt for some sort of meeting-place where science could grow in a sympathetic soil, and be assessed on its own merits.

It was in Italy that such feelings were first translated into action. In 1560 the Accademia Secretorum Naturae was founded in Naples, and a similar society, the Accademia dei Lincei, existed in Rome from 1603 to 1630. Yet a third, the Accademia del Cimento, was founded in Florence in 1657 under the patronage of the Grand Duke Ferdinand di Medici and his brother Leopold, but survived only for ten years.

In England the need for such an organisation had been voiced by Francis Bacon, Lord Verulam, in his *Novum Organum* (1620) and partly, it is thought, because of his writings, Charles II founded the 'Royal Society for the Improvement of Natural Knowledge' in 1662, to provide a meeting ground for English men of science. Actually many of them had already been meeting unofficially and informally from 1645 on—first at Gresham College in London under the name of the Invisible College, then in Oxford during the civil war, and then again in London—so that Charles did little more than set the seal of royal approval on what was already an accomplished fact.

In France the Académie des Sciences was founded by Louis XIV in 1666. There was no corresponding move in Germany until 1700, when the Elector Frederick of Prussia founded the Berlin Academy, although there had been private attempts to found such a society in Rostock as far back as 1619.

These societies had all the same central aim, the increase of natural knowledge by means of free discussion, but their activities took different forms in different countries.

The Italian academies seem to have become deeply involved in the conflicts between science and orthodoxy. The Accademia dei Lincei is said to have supported Galileo in his rebellion against the ecclesiastical authorities (p. 176). The Accademia del Cimento came to an end precisely when Leopold was made a Cardinal of the Church, and many have suspected that the two events were not unconnected; Leopold may have paid for his Hat by the dissolution of a society which had become troublesome to the Church. One of its members, Antonio Oliva, fell into the clutches of the Inquisition, and killed himself to escape torture.*

The English and French academies were mainly concerned with the development of utilitarian science, the study of industrial arts, and the improvement of technical processes. The French Academy expressed this in its title: 'naturae investigandae et perficiendis artibus.' And, although not expressed in its title, this seems to have been the spirit of the Royal Society also; even during its first informal meetings, Boyle had written of 'our new philosophical college that values no knowledge but as it has a tendency to use'. The Society's Royal Patron and his advisers used to direct its attention from time to time to the practical needs of the country. We find him recommending its official experimenter, Robert Hooke (p. 182), to study 'the business of shipping', while Sir Joseph Williamson, Secretary of State, admonished him to be diligent to study things of use.† On the other side of the picture, we read of Charles 'laughing mightily at Gresham College [the embryo Royal Society] for spending time in weighing of air, and doing nothing else since they sat'.‡

This was all in accord with the spirit of the age. Science, which had hitherto been studied mainly for its intellectual interest and for the gratification of curiosity by men who loved it for its own sake, was now recognised as having a definite utilitarian value. Bacon had written much about 'science in the service of humanity', while Boyle had composed a whole treatise on *The Usefulness of Experimental Natural Philosophy* (1663–71), in which he adduced the flourishing trades of spectacle- and clock-making as the fruits of the purely scientific researches of Huygens and Hooke. Even astronomy was coming to be valued for its utility.

* Wolf, *A History of Science, Technology and Philosophy in the Sixteenth and Seventeenth Centuries*, p. 55.

† G. N. Clark, *Science and Social Welfare in the Age of Newton*, p. 15.

‡ Pepys, *Diary*.

Earlier in the century Kepler had remarked that mother astronomy would starve, did not her silly daughter astrology earn bread for both. Now (1675) Charles II founded the Royal Observatory at Greenwich 'in order to the finding out the longitude of places for perfecting navigation and astronomy'. Science no doubt lost much through this shifting of the emphasis from knowledge for its own sake to knowledge for utility's sake, but it also gained much through making a wider and more intelligible appeal to the general mass of the people.

Another favourable influence of a still more material kind was the ever-increasing use of printing (p. 120), which not only brought old knowledge within the reach of all, but also made new knowledge immediately accessible to a wide circle. Every man now stood directly on the shoulders of his predecessors in a way which had been unknown in the past.

Such were the influences which surrounded the science of the seventeenth century and helped it on its way. We must now trace out the history of this progress in detail, again beginning with astronomy, the science in which progress was most spectacular.

ASTRONOMY

Our story of sixteenth-century astronomy ended with Tycho Brahe the Dane, who lived only a short ten months into the seventeenth century. We saw him, shortly before his death, appointing the young German Johannes Kepler as his assistant, and our study of seventeenth-century astronomy may well begin with the life-work of this young assistant.

KEPLER (1571–1630). He was born on 27 December 1571, at Weil near Stuttgart, being the son of a Protestant officer in the service of the Duke of Brunswick. An active brain but enfeebled body—for an attack of smallpox when a child had left him with crippled hands and damaged eyesight—were thought to indicate an ecclesiastical career. Thus he was sent to the Monastic School of Maulbronn and the Protestant

University of Tübingen, where the Professor of Mathematics and Astronomy, Michael Mästlin, convinced him of the truth of the Copernican theories. Kepler now began to feel that his views were too unorthodox for an ecclesiastical career, so he obtained a lectureship in astronomy at Graz in Styria, but abandoned this when a Catholic majority in Graz began to persecute the Protestant minority.

At the early age of 24 he published a book, *Prodromus Dissertationum Cosmographicarum seu Mysterium Cosmographicum*, which contained a reasoned defence of the doctrines of Copernicus, but was even more concerned with the views of its writer. Pythagorean ideas as to the importance of integral numbers were sweeping over intellectual Europe, and Kepler, with a markedly mystical temperament, was especially susceptible to such ideas. Like the Pythagoreans, he felt convinced that God must have created the world after some simple numerical pattern, and, like Plato before him, he tried to discover simple numerical relations between the radii of the planetary orbits. When he failed in this, he began to think that the plan of the universe might perhaps be geometrical rather than arithmetical. His first idea was that the planetary orbits might be spaced like the circles in fig. 26, thus forming inscribed and circumscribed circles to a series of regular polygons. When this idea too proved unworkable, he tried replacing the circles by spheres, and the polygons by the five regular solids of the Pythagoreans (Plate I). His joy was intense when he found no discrepancies too large to be attributed to errors of observation; the world, it seemed, had been built on a simple geometrical pattern, and he declared that he would not barter the glory of his discovery for the whole of the Kingdom of Saxony.

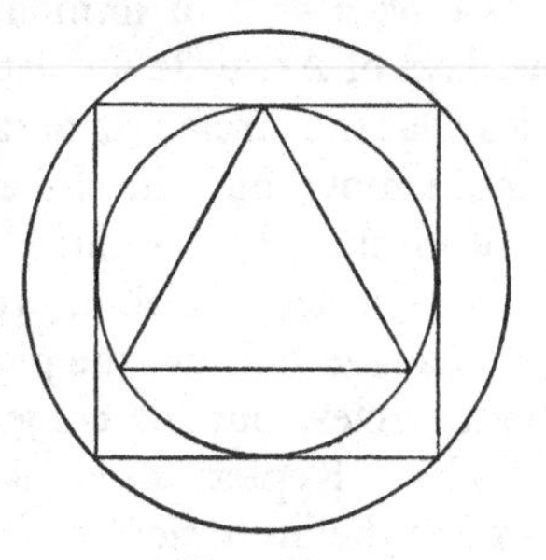

Fig. 26.

It was not so easy to arouse a similar enthusiasm in others. Galileo praised the ingenuity of the book, but left it at that. Tycho offered the chilling advice 'first to try to lay a solid foundation for his views in observation and then, building on this, to try to reach the causes of things'. None the less, Tycho must have appreciated Kepler's intellectual powers, for he invited him to Prague, first as a guest and then as an assistant in the Observatory.

When Kepler accepted this appointment, he was at once set to work on Tycho's vast accumulation of planetary observations. Tycho must have hoped that these would confirm his own anti-Copernican views (p. 135), while Kepler no doubt cherished other hopes. Before the question was settled Tycho died, and Kepler, who succeeded to his position, became free to devise a scheme on Copernican lines which would fit the observations, and so provide a proof of Copernican doctrines. But the best he could do left him with a discrepancy of 8′ of arc in the position of Mars, and this, he rightly judged, was too large to attribute to errors of observation.

He now took a step of immense significance and daring. From the days of Aristotle on, astronomy had been obsessed by the idea that the circle was nature's own curve, so that the planets could move only in circles or in curves which were built up of circles. This reduced astronomy to a perpetual juggling with circles. In about 1080, the Spaniard Arzachel (p. 109) had suggested that the planets might move in ellipses rather than circles, but his conjecture had roused but little interest. Now Kepler also managed to disentangle his thoughts from the limitation to circular motion, and met with the instant reward of a scheme which fitted all the observations perfectly. His book *Astronomia nova* (1609), which announced the results of his labours, enunciated the two laws:

(1) The planet [Mars] moves in an ellipse which has the sun at one of its foci.

(2) The line joining the sun to the planet sweeps out equal areas in equal times.

Nine years later (1618) Kepler published another book, *Epitome Astronomiae Copernicae*, in which he extended these laws to the other planets, to the moon, and to the four newly discovered satellites of Jupiter (p. 173). In his *Harmonices Mundi* (1619), he announced yet another law of planetary motion, now commonly known as Kepler's third law:

(3) The square of the time which any planet takes to complete its orbit is proportional to the cube of its distance from the sun.

These three laws covered all aspects of planetary motion. The first specifies the path in which a planet moves, while the second specifies how it moves on this path, i.e. the way in which the speed varies. As the planet moves nearer to the sun, the sweeping arm—the line joining the sun to the planet—becomes shorter, so that the planet must move faster to sweep out areas at the same rate as before; the nearer a planet is to the sun, the greater its speed in its orbit. The third law informs us as to how the times compare which the different planets require to complete their various orbits. For instance, if one planet *A* is four times as far from the sun as a second planet *B*, then the periodic time of *A* (i.e. *A*'s year) will be eight times as great as that of *B*. But *A* has only four times as far to travel as *B*, so that its average speed will be only half of *B*'s. To take an actual case, Saturn's orbit has 9·54 times the radius of the orbit of the earth. The ratio of the cubes of these distances, namely, 868·3, must be the ratio of the squares of the periodic times (the years) of the two planets. On taking the square root of 868·3, we find that Saturn's year must be equal to 29½ of our years. In general, the nearer a planet is to the sun, the faster it moves.

These three laws of Kepler have been confirmed by innumerable observations. We know now that they are not absolutely exact, but they are so exact that no error was found in them for more than 200 years (p. 299). As they form one of the great landmarks in the history of astronomy, let us pause for a moment to glance back over the road by which they were reached.

The essential part of the story begins with Ptolemy's assuming that sun, moon and planets all moved in orbits round the earth; Ptolemy thought that these orbits were 'epicycles' obtained by superposing one circular orbit on another in the way already explained (p. 94). As the centuries rolled on, and more accurate observations became available, this scheme was seen to be inadequate. Then Copernicus approached the problem with masses of medieval observations, and supposed that the earth moved round the sun. This made a great simplification possible, so that he was able to fit the new observations with only thirty-four circles in place of the previous eighty. Kepler next came with the incomparably better observations of Tycho, and found that they could not be fitted into this scheme of Copernicus, which he now tried to amend. Copernicus had hardly dared to aspire to an accuracy of 10′ of arc, but Kepler rejected his own first attempt because it showed an error of this magnitude. Success came when he tried replacing the age-old circles by ellipses; seven ellipses could now account for the motion of all seven bodies. The new scheme was so simple and so convincing that the Aristotelian circles now dropped out of astronomy for good. Since early Greek days, astronomers had assumed that the planets must move in circles; Kepler had now shown that they did not.

But in answering one question, Kepler had opened up another: Why did the planets move in ellipses rather than in any of the innumerable other curves that could be imagined? In the past this question of 'why' could be answered by saying that the circle was nature's own curve; now that nature had repudiated the circle, another type of answer seemed to be called for. The question was to dominate astronomy until Newton appeared and gave an answer which was not final but, like Kepler's answers to the question of 'how', seemed to be final for about 200 years.

Kepler himself gave some thought to the question of why, but without any great success. Following Pythagoras, he had thought in turn that the clue to the planetary orbits was to be found in numbers, in polygons, and in solids. He now turned to music. Using the Pythagorean relation between the lengths of strings and the musical notes they produced, he represented the motions of the planets in their orbits by a group of musical notes—the harmony of the spheres (p. 60)—and the same *Harmonices Mundi* which announced Kepler's third law of planetary motions also announced the chords which the planets must sing as they moved in their paths round the sun.

But this led nowhere, and Kepler next turned to mechanics. He still held the Aristotelian view that if a body was in motion, something must be pushing it on from behind. So he assumed that the sun was occupied by an 'anima motrix' (motive force) which did the pushing. It emitted tentacle-like rays of force which rotated with it,* like the spokes of a wheel, and urged the planets on. In brief, Kepler replaced Aristotle's rotating 'sphere of the fixed stars' by a rotating sun.

This might explain circular orbits, but could not of itself explain elliptical orbits; something more was needed for these. Gilbert had suggested that the earth attracted the moon like a magnet, and Kepler now pictured every planet as a magnet of which the axis always pointed in the same direction, this direction lying in the plane in which the planet moved round the sun. Thus as a planet moved in its orbit, it was alternately attracted and repelled by the sun, and Kepler thought that this would cause alternate increases and decreases in the distance of the planet from the sun—hence the elliptic orbits.

Kepler thought that similar forces might operate throughout the universe, and to this extent he toyed with the idea of universal gravitation, which he described as 'a mutual affection between bodies, tending towards union or conjunction, and similar in kind to magnetism'. In support, he instanced the tendency of all bodies to fall earthwards—'the earth

* There was so far no proof that the sun rotated.

attracts a stone rather than a stone seeks the earth'—and the tides of the ocean which, like Gilbert and many others, he attributed to the moon's attraction. He said that two stones in empty space would attract one another, and would finally meet at the point which we now call the centre of gravity of the two. If the moon and the earth were not held in their orbits by their 'anima motrix' or its equivalent, the earth would rise to the moon through a fifty-fourth part of their distance apart, while the moon would fall to the earth through the other fifty-three parts. But Kepler never suggested, and apparently never suspected, that this same gravity could keep the moon and planets in their orbits without calling in an 'anima motrix' to supply a push.

Kepler's second law had shown him that the nearer a planet is to the sun, the faster it moves, and he interpreted this as showing that the 'anima motrix' must be greatest at small distances. He conjectured that its strength must fall off as the inverse square of the distance, so that at double distance its force would be only a quarter as strong. But he soon abandoned this opinion, and decided that the force fell off as the inverse distance, so that at double distance it would be half as strong. The French astronomer Bouilliaud disagreed with this change, and argued for Kepler's original law of the inverse square,* and here the question stood until Newton appeared on the scene—to show that the true law was that of the inverse square, but that no 'anima motrix' was needed beyond the sun's gravitational pull.

Early Telescopic Astronomy

Meanwhile observational astronomy did not stand idle. In the same year in which Kepler published his *Astronomia nova* and the first two laws of planetary motion, Galileo made his first telescope.

The first telescopes. The ultimate origin of the telescope is hidden in some obscurity. Roger Bacon had vaguely ex-

* *Astronomia Philolaica*, Paris, 1645.

plained the principles on which such an instrument might be constructed (p. 116), but we do not hear of his ever trying to make one. An English mathematician, Leonard Digges of Oxford, is reported to have done so, and his son wrote of his using it, but no astronomical discoveries or further activity of any kind seems to have followed,* and the practical invention of the telescope, like that of the microscope, must be attributed to the Dutch spectacle-makers. Early in the seventeenth century they were making telescopic instruments of the opera-glass type, but treating them only as toys. Official records at the Hague show that a patent for the making of these instruments was granted to a certain Hans Lippershey of Middelburg on 2 October 1608, and that a similar application from a James Metius was considered on 17 October. Descartes credits the invention of the telescope to Metius, but Lippershey is generally supposed to have the better claim. Like Jansen's invention of the microscope, Lippershey's discovery is said to have been largely accidental; one day he happened to turn a combination of lenses towards a distant weather-vane, and was surprised to see it substantially magnified.

As soon as Galileo heard of this, he saw its scientific importance, and set about making such an optical glass for himself.† By 'sparing neither labour nor expense', he had soon constructed 'an excellent instrument' which magnified objects about a thousand times in area, and so reduced their apparent distances to about a thirtieth. In the next year Kepler suggested a better arrangement of lenses,‡ and an instrument embodying the improvement was constructed by the Jesuit Scheiner. A few years later, Huygens (p. 210) made still further improvements which brought the instrument almost to its present form.

GALILEO. Hitherto Galileo had taken no great part in astronomy, but he seems to have been sympathetic to

* Wolf, loc. cit. p. 75.

† *Sidereus Nuncius.*

‡ *Dioptrice*, 1611.

Copernican views, having been persuaded of their truth, some say, by that same Mästlin who had converted Kepler. But as soon as he had a telescope in his hands, he proceeded to make discovery after discovery at breakneck speed. Turning it on to the moon, he saw markings which he interpreted as shadows cast by surface irregularities. 'We have appearances quite similar on the Earth about sunrise, when the valleys are not yet flooded with light, but the mountains surrounding them on the side opposite to the sun are already ablaze with the splendour of his beams; and just as the shadows in the hollows of the earth diminish in size as the sun rises higher, so these spots on the moon lose their blackness as the illuminated part grows larger and larger.' The moon, then, was a world like our own, and the Aristotelians, who had asserted that it must be a perfect sphere, had been wrong.

He next turned his telescope on to well-known areas of the sky and saw vast multitudes of new stars. 'Beyond the stars of the sixth magnitude* you will behold through the telescope a host of other stars, so numerous as to be almost beyond belief.'† For instance, the Belt and Sword of Orion no longer contained only nine stars but more than eighty, while the six well-known stars of the Pleiades were increased to thirty-six and more. The effect was even more marked in the Milky Way. 'Upon whatever part of it you direct the telescope, straightway a vast crowd of stars presents itself to view; many of them are tolerably large and extremely bright, but the number of small ones is quite beyond determination.' 'The galaxy is nothing else but a mass of innumerable stars planted together in clusters', which is precisely what Anaxagoras and Democritus had said 2000 years earlier.‡ Bruno, too, had been right or nearly right; the number of stars was certainly very great in number, and might well be infinite. Galileo next

* According to the classification of Hipparchus these are the faintest stars which are visible to the naked eye, and so the faintest which could be seen at all in pre-telescopic times.

† *Sidereal Messenger*, 1610.

‡ Aristotle, *Meteorologica*.

turned the telescope on to Jupiter, and saw four satellites circling round the great planet, just as Copernicus had imagined the planets to circle round the still greater mass of the sun. Probably, then, Copernicus had also been right, since here was a miniature replica of the solar system as he had imagined it to be. A study of Venus showed that it passed through a sequence of 'phases' like those of the moon—crescent, semicircle, full circle, semicircle, crescent again, and so on indefinitely. This proved that Venus was not self-luminous, but shone by light reflected from the sun. But it showed more than this. For the Ptolemaic hypothesis required that Venus should never show more than a semicircle of illuminated surface to the earth, whereas the Copernican hypothesis demanded exactly the sequence of phases that were actually seen. Thus this one observation of Galileo killed the Ptolemaic hypothesis and established the Copernican for all who could be convinced by visual proof. On 30 January 1610, Galileo wrote: 'I am quite beside myself with wonder, and infinitely grateful to God that it has pleased Him to permit me to discover such great marvels.'

Later in the same year he observed the rings of Saturn, but interpreted them wrongly, writing that Saturn consisted of 'three spheres that almost touch one another'. He also made observations on sunspots. Kepler had observed a dark spot on the sun in 1607 without telescopic aid, but had thought it was Mercury crossing the disk of the sun. Fabricius also observed sunspots and the Jesuit Scheiner observed some in April 1611. At first Galileo thought they were an optical illusion but later, when he became convinced of their reality, he saw that they proved that the sun must be rotating, and that they gave a means of determining the period of this rotation.

In 1613 Galileo published his *Letters on the Solar Spots* in which he made no secret of his Copernican views, and in consequence found himself charged with heresy. He tried to defend himself by quoting scripture, but the ecclesiastical authorities warned him that he must abstain from theological argument and confine himself to physical reasoning. Early in 1616 the matter was brought before the Inquisition, who summoned a meeting of their consulting theologians on 19 February to advise on the two propositions:

(1) The sun is the centre of the world and altogether stationary.

(2) The earth is not the centre of the world nor stationary, but moves bodily with a diurnal motion.

The meeting unanimously pronounced the first proposition to be 'false and absurd in philosophy and formally heretical', and adjudged the second proposition 'to deserve the like censure in philosophy, and as regards theological truth to be at least erroneous in faith'. On 25 February the Pope directed Cardinal Bellarmine, the leading member of the Inquisition, to summon Galileo and admonish him to abandon the opinion which the meeting had condemned. If Galileo refused, he was to be handed a formal injunction ordering him, under pain of imprisonment, 'to abstain from teaching or defending this kind of doctrine and opinion, or from treating of it'. Galileo was duly summoned and saw Bellarmine the next day —with what precise results we do not know. But a decree appeared a week later ordering the work of Copernicus to be withdrawn from circulation until it had been corrected. The book reappeared four years later, with 'corrections' on the lines of those which Osiander had previously made—suggestions that the earth's motion was not absolute truth, but only a hypothesis which facilitated calculations.

After a period of quiescence, Galileo published *Il Saggiatore* ('The Assayer') in 1623, dedicating it to the new Pope Urban VIII, who had been sufficiently sympathetic with astronomy to write a poem celebrating Galileo's discovery of the satellites of Jupiter. He was apparently willing to turn a blind eye to the various unorthodox tendencies of the book, and the rift between religion and science, at least as represented by the Pope and Galileo, might have been healed if only Galileo could have let sleeping dogs lie. But it was not in his temperament to do this, and in January 1632 he published a *Dialogue on the Two Chief World-Systems, the Ptolemaic and the Copernican*. Then the lightning struck.

In the book three characters meet to discuss the merits of the two systems. One is, of course, a convinced Copernican, another a violent anti-Copernican, steeped in the doctrines of Aristotle, while the third claims to be an impartial onlooker and commentator. Impartial comment was not in any case Galileo's strong suit, and least of all here, where he could hardly see that there was any case beyond that of Copernicus to comment on. And he made the anti-Copernican desperately stupid, and incapable of seeing the simplest of arguments.

Galileo's writing of such a book was of course an open defiance of the Pope's admonition of 25 February 1616 to abandon Copernican opinions, and even more so of the formal injunction of 26 February, if this was ever handed to him.* Nevertheless, Galileo had succeeded in obtaining a permit for its publication from the Censor of the Holy Office, subject to two conditions. The first was the old condition that the motion of the earth was to be treated as a hypothesis and not as a fact; the second was that the book should contain certain arguments, to be supplied by the Pope, in favour of the orthodox view. Galileo not only failed to comply with the first condition but also, quite gratuitously, laid himself open

* Some authorities have maintained that the injunction was not handed to Galileo, and even that the whole episode of the admonition and injunction is a fabrication, trumped up at a later date to provide an excuse for the proceedings of 1633 (see especially Emil Wohlwill, *Der Inquisitionsprocess des Galileo Galilei*, Berlin, 1870). Galileo's own statement was: 'There was published some years since in Rome a salutiferous Edict that, for the obviating of the dangerous scandals of the present age, imposed a seasonable Silence upon the Pythagorean Opinion of the Mobility of the Earth....I was at that time in Rome; and not only had the Audiences but Applause of the most eminent Prelates of that Court; nor was that Decree published without previous Notice given me thereof' (Galileo, *Dialogue on the Two Chief World-Systems*, translation by Salusbury, 1661).

to further trouble by the way in which he complied with the second. For he put the Pope's arguments into the mouth of his very stupid anti-Copernican character.

The sunspot-observing Jesuit Scheiner, with whom Galileo was already on bad terms, owing to a dispute as to priority in the discovery of sunspots, seized this opportunity of making mischief, and asserted that the half-wit was intended for a pen-portrait of His Holiness. In August 1632 the sale of Galileo's book was forbidden, and a commission was appointed to report upon it. They reported unfavourably, and Galileo was summoned to appear before the Inquisition.

He arrived in February 1633, and was at once put under detention, although he appears to have been treated with great consideration and with every courtesy. Two months later he was examined and is said to have been threatened with torture, although it is also said that there was no intention of putting the threat into execution. In June sentence was pronounced—Galileo was to do penance for three years and to make a recantation in penitential garb of all Copernican doctrines. In this recantation he was not only to state his present beliefs, but also to predict what he would believe for the rest of his life: 'I, ..., give assurance that I believe, and always will believe, what the Church recognises and teaches as true.' There is no reason for accepting the highly improbable story that Galileo ended his recantation with the muttered words 'E pur si muove' ('And yet it moves'). It is the kind of thing Galileo would have said, but hardly the kind of occasion he would have chosen for saying it.*

After another period of detention, he was allowed to move to Arcetri, outside Florence, where he continued to work on scientific problems, although these were now of a non-controversial kind. Here in 1637 he discovered the 'librations' of the moon—the small oscillations of the moon which result

* So far as is known, this story first appeared in 1761, in the Abbé Irailli's *Querelles littéraires*.

in its showing slightly different parts of its surfaces to the earth in successive lunar months.

He also gave thought to methods for enabling a ship to determine its position when out of sight of land. The problem had not arisen for ancient or medieval navigation, because this had consisted of little more than 'hugging the coast', but with the discovery of the New World other methods were called for. A ship can find its latitude very easily by noting the greatest elevation reached by the sun, but the determination of its longitude presents a far more difficult problem. In 1598 Philip III of Spain offered a prize of 100,000 crowns for a method of fixing a ship's position when it was out of sight of land, and the Dutch soon followed it with a similar offer.

There is a solution which is simple enough in principle. When we say that a ship's position is 60° west of Greenwich, we mean that noon occurs there four hours later than at Greenwich.* If, then, a clock showing Greenwich time were available on the ship, the longitude of its position could be found at once by noting the time which this clock showed at local noon, i.e. when the sun was at its highest. The modern ship must carry chronometers which show Greenwich time throughout the voyage, and may receive Greenwich time by radio signals, but no such methods were available to seventeenth-century navigators, who were often 200–300 miles from their estimated positions, and frequently lost their ships as a result.†

Galileo pointed out that time-tables of the motion and eclipses of Jupiter's satellites could be prepared in advance, and that to a navigator provided with such tables, the system of Jupiter would provide the needed clock from which a standard time could be read and the longitude of a ship's position determined. The method was hardly used, because it was soon found that the motion of the moon across

* Because 60° is a sixth of a circle, and four hours is a sixth of a day, the position is one-sixth of the way round the world from Greenwich.

† Sir H. Spencer Jones, *The Royal Observatory, Greenwich*, 1944.

the sky provided a still better clock; the mariner can tell the standard time at any moment by noting the position of the moon relative to the nearest stars—in principle it is as simple as reading the hands of a watch.

Galileo became blind in 1637—as some think from observing the sun without adequate protection for his eyes—and his life's work was at an end. He died in 1642, the year of the birth of Newton, who was to extend and give a wider significance to his labours.

Just as Galileo's earlier work in mechanics had shattered the Aristotelian physics, so his later work in astronomy had shattered the Aristotelian cosmology. By the beginning of 1610 he had observed the phases of Venus, the orbital motion of Jupiter's satellites, and the uncountable stars of the Milky Way, while Kepler had enunciated his first two laws of planetary motion. The main scheme of the universe was now clearly established, and it was obvious that the final victory could only rest with Copernicus, Bruno and Galileo.

Descartes' Vortices

DESCARTES (1596–1650). The next important step—important though retrograde—was taken by Réné Descartes the philosopher, whom we have already met as one of the founders of analytical geometry. He was born near Tours of a good family, was educated at the Jesuit school of La Flèche where he showed himself exceptionally brilliant at mathematics, and then studied mathematics for two years with Mersenne, the famous Parisian mathematician. After a brief period in the army of Prince Maurice of Orange, he resigned his commission at the age of 25 to devote the rest of his life to mathematics and philosophy.*

* While he was stationed at Breda, he noticed a placard in the street written in Dutch, and asked a passer-by to translate it to him. It proved to contain a mathematical problem and a challenge to all the mathematicians of the world to solve it. The stranger, who was Beeckman, principal of the college of Dort, had said he would provide the translation if Descartes would undertake to provide the solution. This Descartes did

V

Guericke's experiments with Magdeburg hemispheres

After a time of travel, he settled down in Holland, and worked for five years on a book *Le Monde*, which aspired to give a complete outline of science as well as a complete theory of the physical universe; he was still young and was not a modest man. As this book was approaching completion in 1633, he heard of Galileo's condemnation. He explained* that, although he had noticed nothing in Galileo's doctrines which had seemed to him prejudicial either to religion or to the state, he now began to feel anxious about some of his own doctrines, and so decided not to publish his book. Actually others published it, still in an unfinished state, in 1664—14 years after his death. But he gave a brief account of his conclusions in his *Discours de la Méthode* (1637), and a fuller exposition of his views in his *Principia Philosophiae* (1644).

These conclusions were valueless but Descartes' theory is of interest as being the first attempt to explain the universe on purely mechanical lines. It is the theory of a philosopher rather than of a scientist, being based on general principles, contemplation and conjecture rather than on experiment. For instance, he condemns Galileo's experiments with the comment that 'everything that Galileo says about the philosophy of bodies falling in empty space is built without foundation; he ought first to have determined the nature of weight'.

He followed Galileo in dividing the qualities of substances into two groups, which he called primary and secondary. The secondary qualities are the hardness or softness, the sweetness or acidity, and so on, that need senses to perceive them (or so Descartes thought), while the primary qualities are those which exist in their own right, whether they are perceived or not. Descartes says there are only two primary qualities—extension in space and motion—so that nothing has any real

within a few hours, and the evidence of his mathematical powers is said to have encouraged him to return to the mathematical pursuits of his earlier days.

* *Discours de la Méthode*, Part VI, 1637.

objective significance except these: 'give me extension and motion, and I will construct the world.'

He proceeds to argue that, as extension is the fundamental property of matter, extension without matter is unthinkable —a strange argument for a philosopher who claimed to accept nothing that could not be established with certainty.* Thus all space must be occupied by matter of some kind or other; 'a vacuum or space in which there is absolutely no body is repugnant to reason'. He accordingly imagines that all those parts of space which are not occupied by the solid matter of our experience are occupied by other 'primary' matter consisting of very fine particles which make no impression on our senses.

When a fish swims through the sea it pushes particles of water away from in front of itself, while other particles close in from behind to fill the gap vacated by the fish, so that the water moves round and round in closed circuits. 'All natural motions are in some way circular.' In the same way, Descartes thought, when ordinary gross matter pushes its way through the sea of particles, this must move in closed circuits, and so may be pictured as a series of vortices.

On this foundation, Descartes built his famous theory of vortices. The vortices were whirlpools in a sea of particles; ordinary material objects were like floating corks which revealed how the currents were flowing in the whirlpools. The finest particles of all, which were rubbings or filings from the coarser kinds, were drawn towards the centres of the vortices. The planets were corks caught in the whirlpool of the sun and whirled round its centre, while a falling leaf was a smaller cork being drawn towards the centre of the earth's whirlpool. In a later elaboration, there was supposed to be so much agitation at the centre of a large whirlpool that objects became luminous; this explained why the sun and the stars shone.

* *Principia Philosophiae*, Part II, § 15. A similar argument would be that as motion is the fundamental property of a locomotive, *ergo* motion without a locomotive is unthinkable.

This system attained to a vogue out of all proportion to its scientific merits, partly no doubt because it was easy to visualise and so seemed easy to understand. But it made no attempt to explain quantitative laws, such as Kepler's laws of motion, and showed no capacity for surviving such tests when applied by others—as Newton very conclusively demonstrated.* Nevertheless, this theory of Descartes held the field, and was indeed the best that science had to offer, until it was superseded by the incomparably better theory of universal gravitation, to which we now turn.

Universal Gravitation

We have seen (p. 169) how Kepler toyed with the idea of universal gravitation, but had no suspicion that forces of gravity alone might explain the motions of the planets; indeed, he thought that these could not maintain their orbital motions unless some force was continually pushing them on from behind. Actually, of course, a planet's motion is maintained by its own momentum. What was needed to explain the observed facts was not a pushing force to keep the planet continually moving on, but an attractive force which would continually change the direction of motion of the planet, and so prevent it running away in a straight line from the sun.

The general principle involved had been stated very clearly by Plutarch 1400 years before Kepler was born, although with special reference to the moon's motion round the earth. He had written: 'The moon is secured against falling [on to the earth] by her motion and the swing of her revolution—just as objects put in slings are prevented from falling by the circular whirl'†—and it was only necessary to discover what played the part of the sling.

In 1666 no fewer than three people took up the problem almost simultaneously; in alphabetical order they were Borelli, Hooke and Newton.

* *Principia*, II.

† *De Facie in Orbe Lunae.*

BORELLI. Borelli, who was Professor of Mathematics in Galileo's old University of Pisa, published a book* in which he said that a planet moving in a circular orbit round the sun had a tendency to recede from the sun and, like Plutarch before him, he compared the motion with that of a stone in a sling. He argued that as a planet does not in actual fact recede from the sun, there must be some force drawing it towards the sun; when the attractive tendency of this force is just equal to the recessive tendency caused by the motion, then equilibrium is established, and the planet will revolve continually round the sun at a definite distance from it. It was the first time that the mechanics of the problem had been accurately stated since the time of Plutarch.

HOOKE (1635–1703). The same thing was being said in England at almost the same time by Robert Hooke, who was both an acute thinker and an ingenious experimenter. After being employed as research assistant from 1655 to 1662 by Robert Boyle, whom we shall meet later (p. 213), he was engaged as Curator of the newly founded Royal Society, his duty being to carry out experiments suggested either by his own fertile brain or by the other Fellows. In a paper dated 23 May 1666, he discussed how the path of a celestial body could be bent into a circle or ellipse, and considered that it might be through 'an attractive property of the body placed in the centre [of the orbit] whereby it continually endeavours to attract or draw it [i.e. the celestial body] to itself'. Hooke says that if such a force is granted, then 'all the phenomena of the planets seem possible to be explained by the common principles of mechanic motion'.

In another paper published eight years later, he attempted to 'explain a system of the world differing in many particulars from any yet known, but answering in all things to the common rules of mechanical motions. This depends upon three

* *Theoricae Mediceorum Planetarum* (1666). This dealt primarily with the motion of Jupiter's satellites, which Galileo had called the Medicean stars.

suppositions.' The first supposition is simply universal gravitation. 'The second is that all bodies whatsoever continue to move forward in a straight line until they are deflected by some effectual powers, and sent into a circle, ellipse, or other more complicated curve. The third supposition is that these forces are strongest at short distances, and fall off in strength as the distance is increased.'

Here Hooke accurately enunciates the mechanical principles which govern the motions of the planets, and suggests a universal force of gravitation. He does not say how the force must vary with the distance if it is to result in the planets moving in the ellipses that are actually observed. Five year later (1679) he wrote to Newton that if the force varied as the inverse square of the distance, then the orbit of an object projected from the earth's surface would be an ellipse, having one focus at the centre of the earth.* The theory of planetary motions and of universal gravitation was now almost complete, but it needed the genius of Newton to weld it into a consistent whole, and to establish that the mysterious force that kept the planets moving in their orbits round the sun was identical with the familiar force which caused an apple to fall to the earth.

NEWTON (1642–1727). Isaac Newton was born prematurely on Christmas Day, 1642, at the Manor House of Woolsthorpe, near Grantham, in Lincolnshire, being the posthumous son of a yeoman farmer who was Lord of the Manor of Woolsthorpe. He was so small at birth that his mother said she could have put him in a quart pot, and so lacking in vitality that the two women who went to fetch a 'tonic medicine' for the poor child were surprised to find him still alive when they returned.

In due course he was sent to school at Grantham. He did not make a good pupil, being inattentive, according to his own statement, and standing low in his class. But he showed a certain mechanical aptitude in his play, devising ingenious

* Brewster's *Memoirs of Sir Isaac Newton*, I, 287.

ways of measuring the speed of the wind, making clocks, sundials and a working model of a windmill, and designing a carriage which was to go when its occupant turned a handle.

When he was 13 his mother lost her second husband, and Newton was called home to help on the farm. But it soon became clear that his interests lay elsewhere, and he gave more thought to mechanical problems than to agriculture. Finally, it was decided that he would never make a good farmer, and so he might try to be a scholar. He was sent to Trinity College, Cambridge, the college where his uncle had been educated, probably entering in June 1661.

There is no evidence that he was specially interested in science at this time, or that he specially impressed the College authorities by his ability. A book on astrology seems to have been more effective than they in awakening his interest in science. For in it he found a geometrical diagram which he could not understand, and he then bought a 'Euclid' and set to work to learn geometry. Having mastered this book with ease, he went on to study the far more difficult 'Geometry' of Descartes, which seems to have given him a real interest in mathematics and a taste for science.

In the summers of 1665 and 1666 England was ravaged by the plague, and the Cambridge students were sent home to avoid infection. In the quiet of Woolsthorpe, Newton found leisure to ponder over many of the scientific problems of the day, and made good progress towards the solutions of many of them. Writing of this some 50 years later,* he says: 'In the beginning of the year 1665, I found the method for approximating series, and the rule for reducing any dignity [i.e. any power] of any binomial to such a series.† The same year in May I found the method of tangents of Gregory and Slusius,

* In a MS. quoted in the preface to *A Catalogue of the Newton MSS., Portsmouth Collection* (Cambridge, 1888), p. xviii. This MS. was probably written in or about the year 1716.

† This is the famous 'binomial theorem', see p. 225 below.

and in November had the direct method of fluxions,* and the next year in January had the theory of colours, and in May following, I had entrance into the inverse method of fluxions [i.e. the integral calculus], and in the same year [1666] I began to think of gravity extending to the orb of the moon, and... from Kepler's Rule of the periodic times of the planets, I deduced that the forces which keep the planets in their orbits must be reciprocally as the squares of their distances from the centres about which they revolve; and thereby compared the force requisite to keep the moon in her orbit with the force of gravity at the surface of the earth, and found them answer pretty nearly. All this was in the two plague years of 1665 and 1666, for in those days I was in the prime of my age for invention, and minded mathematics and philosophy more than at any time since.' Thus before he was 24 years old, he had already thought out a programme for a large part of his life's work.

Returning to Cambridge, he was elected a Fellow of his College in 1667. Two years later, the then Lucasian Professor of Mathematics in the University, Isaac Barrow (1630–77), who was himself no mean mathematician, resigned his chair for the express purpose of making a vacancy for Newton. When Newton was duly elected, he became free to devote the whole of his time to science.

He worked quietly in Cambridge until 1689, when he was elected Member of Parliament for the University. But this particular Parliament only lasted for 13 months, so that the next year he was back in Cambridge. In 1696 he moved permanently to London, having been appointed Warden of the Mint; he was promoted to the Mastership three years later.

All accounts agree that Newton made a highly competent Master of the Mint, and it was largely through his efforts that the English currency was put on a satisfactory basis at a

* This is the differential calculus, the most famous and the most important of Newton's mathematical discoveries.

difficult time. He noticed a relation between prices and the amount of money in circulation, which was subsequently formalised in the so-called 'quantity theory' of money—if the amount of currency in circulation is doubled, other things remaining the same, then prices also (in terms of currency) will approximately double. It is a simple application in the economic field of the well-known principle that it is impossible to get something for nothing, but apparently it needed a Newton to find it. There is an obvious comparison with Copernicus, who advised the Polish government on currency questions (p. 125), and in doing so discovered another important currency law, which is usually known as Gresham's Law: 'When bad money is recognised as legal tender, good money is driven out of circulation.' Copernicus seems clearly to have anticipated Gresham in the formulation of this law, but it is equally clear that Oresme (p. 118) anticipated Copernicus. It is strange that so many of the world's astronomers should have been intimately mixed up with the problems of filthy lucre.

Newton's acceptance of office in the Mint virtually ended his original work in science, but he was elected President of the Royal Society in 1703, and every year afterwards until his death. He died on 20 March 1727, after two or three years of failing health.

The quotation given above (p. 185) shows that he had thought out the main lines of his famous theory of gravitation by 1666. Others, as we have seen, were thinking or had thought along the same lines, but Newton did one thing which others had not done; he applied a numerical test, and found that the result came out 'pretty nearly' right.

The principle of this test is simple. Representing the orbit as approximately circular, then if the sun's attractive force were suddenly to fail when a planet was at P (fig. 27), the planet would no longer follow the circle PR, but would begin to move along the straight line PQ, and after a second of time would reach some point Q on this line. Actually the sun's force pulls it down to some point R on the circular orbit, so that in one second it falls sunward through a distance QR. Now suppose that the orbit of this planet is four times the size of the orbit of the earth. Kepler's third law (p. 167) tells us that the planet will take eight times as long as the earth to get round its orbit, so that the distance PQ which it covers in one second will be half as great for the earth. But QT, which is approximately the diameter of the orbit, is four times as big as for the earth, so that QR, which is known to be equal to PQ^2/QT, will be one-sixteenth as great as for the earth. Thus the pull of the sun on a planet which is at four times the earth's distance, must be only a sixteenth of the pull at the earth's distance. It can be shown in the same way that the pull at n times the earth's distance will be $1/n^2$ times the pull at the distance of the earth. This is the famous law of the inverse square.

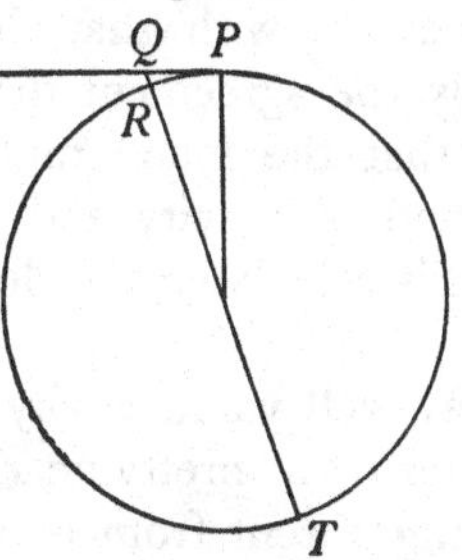

Fig. 27.

Newton might have enunciated this law, and brought forward every planet which conformed to Kepler's law—as they all did—in confirmation. But he did not do this because he wanted to go farther and show that the force of gravitation which kept the planets in their orbits was identical with the familiar 'gravity' which caused a stone or apple to fall to the ground. His niece Catharine Barton told both Voltaire and Martin Folkes, the President of the Royal Society, that it was the fall of an apple in the orchard at Woolsthorpe that first put this train of ideas into Newton's mind, while Newton's friend Stukeley records that he was told by Newton himself that the fall of an apple had first put the notion of gravitation into his mind.*

Newton knew that the earth's gravitation caused terrestrial

* *Stukeley's Memoirs of Sir Isaac Newton's Life* (ed. by A. H. White, 1936), p. 19.

objects to fall 16 feet towards the earth's centre in one second. If Newton's conjecture were sound, the moon, being 60 times as distant from the earth's centre, would fall only $\frac{1}{3600}$th part of 16 feet in a second. Newton tried to calculate how far the moon actually did fall, and found that the result agreed 'pretty nearly' with that demanded by his theory. But apparently the agreement did not satisfy him, and he conjectured that the moon might be kept in her orbit by a combination of gravity and the vortices of Descartes. He now put his calculations aside and did not return to them for 20 years.

We may well wonder why Newton's calculations did not agree better than 'pretty nearly'. His friend Pemberton says that, 'being absent from books' at Woolsthorpe, he used a wrong value for the radius of the earth, which would of course give a wrong value for the 60-times larger orbit of the moon. He may, as Pemberton thinks, have confused English miles with nautical miles; or he may have taken a wrong value from a book he is known to have possessed, which gave the degree as only 66 miles. Yet another alternative is that he was not sure what the earth's gravitational attraction would be on a nearby object such as a falling apple. In 1685 he proved that it would be the same as though all the earth's substance were concentrated at its centre; knowing this, and now using a correct value for the radius of the earth, he found that his calculations agreed perfectly with observation. But in 1666 he could obtain no such agreement, and was content simply to let the matter drop. Perhaps his fertile mind had become more interested in some other problem.

Near the end of 1679 he wrote to Hooke saying, among other things, that he had for some years been trying to 'bend himself from philosophy to other studies' because he grudged the time this took 'except perhaps at idle hours sometimes as a diversion'. We do not know what was now claiming his attention. It may perhaps have been theology, a taste for which grew upon him with advancing years.* Or it may have been chemistry, to which he gave an amount of time out of all proportion to his achievements in the subject.†

In this letter he remarked that if the earth were rotating, a falling object would be carried a little eastward in the course of its fall, and suggested that a proof of the earth's rotation might be obtained by dropping a body from a height and noting where it struck the ground. Early in 1680 Hooke replied that he had performed the experiment and had obtained the expected result, and then went on to ask Newton what would be the exact path of a falling object which was attracted to the earth by a force which varied as the inverse

* Newton's tastes and interests all seem to have been of an intensely serious kind. He must have been quite devoid of humour and seldom laughed or joked. In the only known instance of his laughing (recorded twice, see Brewster's *Memoirs*, II, 91) Newton did not laugh *with* a friend, but *at* him—because he could not see the use of learning Euclid. The catalogues of Newton's books and furniture made after his death suggest that he had no pleasure or interest in art, music, literature or poetry (Villamil, *Newton the Man*), and he appears to have been equally uninterested in country life, animals, exercise, sports and games, both indoor and outdoor. He was ultra-careless about his dress and food. He never married, and one boy-and-girl affair seems to have been the sum total of his interest in the other sex. It may be that his intense intellectual efforts left him with but little mental energy for more human interests, and often with but little even for his science.

† His amanuensis, Humphrey Newton, wrote (Brewster's *Memoirs*, II, 93) that 'at spring and the fall of the leaf...he used to employ about six weeks in his laboratory, the fire scarcely going out either day or night, he sitting up one night, and I another, till he had finished his chemical experiments, in the performances of which he was the most accurate, strict, exact. What his aim might be, I was not able to penetrate into, but his pains, his diligence at these set times made me think he aimed at something beyond the reach of human art and industry.' 'He would sometimes, tho' very seldom, look into an old mouldy book which lay in his laboratory, I think it was titled *Agricola de Metallis*, the transmuting of metals being his chief design.' All the time and energy which Newton gave to these subjects led only to insignificant results: he published nothing original in chemistry, and had no success that we know of in his alchemy—if he really was as interested in these questions, as his amanuensis seems to have thought. (For a discussion of Newton's work and interest in chemistry, see Douglas McKie, 'Some notes on Newton's chemical philosophy', *Phil. Mag*. Dec. 1942.)

square of the distance. Newton says, in a letter written to Halley in 1686, that he looked into the problem to satisfy his own curiosity, and found that the path would be an ellipse having the centre of attraction at one of its foci—exactly the type of path in which Kepler had found the planets to move round the sun. He then 'threw the calculation by, being upon other studies', and left Hooke's letter unanswered.

Four years later (January 1684), Hooke, Halley the astronomer (p. 195) and Sir Christopher Wren, the astronomer and architect, met in London. They had all reached the conclusion, although for different reasons, that the true law of gravitation must be that of the inverse square, a conclusion which Newton had reached as far back as 1666 (p. 185), as a deduction from Kepler's third law, but a further test remained; if the planets moved round the sun under the attraction of such a force, would they move in ellipses, as Kepler first asserted that they did? Halley undertook to go to Cambridge and consult Newton on the question. When they met, Newton at once said the paths would be ellipses, and explained that he had worked the problem out some years before but had mislaid his calculations; for the second time, then, he had held the solution of a large part of the problem of the astronomical universe in his hands, and had been content to let it slip out. This time, however, he promised to restore the lost calculations, and not only did this but also, urged by Halley, he wrote out his results for the Royal Society.

In due course the manuscript was published under the title *Philosophiae Naturalis Principia Mathematica*, which is generally abbreviated to the *Principia*. This is certainly the greatest scientific work ever produced by the human intellect; no other, with the possible exception of Darwin's *Origin of Species*, has had so great an influence on contemporary thought. For it explained a large part of inanimate nature in mechanical terms, and suggested that the remainder might admit of explanation in a similar way. The key to the explanation was of course the law of universal gravita-

[39]

BS, & in loco *C* secundum lineam ipsi *dD* parallelam, hoc est secundum lineam *CS*, &c. Agit ergo semper secundum lineas tendentes ad punctum illud immobile S. *Q. E. D.*

Cas. 2. Et, per Legum Corollarium quintum, perinde est sive quiescat superficies in qua corpus describit figuram curvilineam, sive moveatur eadem una cum corpore, figura descripta & puncto suo S uniformiter in directum.

Scholium.

Urgeri potest corpus a vi centripeta composita ex pluribus viribus In hoc casu sensus Propositionis est, quod vis illa quæ ex omnibus componitur, tendit ad punctum S. Porro si vis aliqua agat secundum lineam superficiei descriptæ perpendicularem, hæc faciet corpus deflectere a plano sui motus, sed quantitatem superficiei descriptæ nec augebit nec minuet, & propterea in compositione virium negligenda est.

Prop. III. Theor. III.

Corpus omne quod, radio ad centrum corporis alterius utcunq; moti ducto, describit areas circa centrum illud temporibus proportionales, urgetur vi composita ex vi centripeta tendente ad corpus alterum & ex vi omni acceleratrice, qua corpus alterum urgetur.

Nam (per Legum Corol. 6.) si vi nova, quæ æqualis & contraria sit illi qua corpus alterum urgetur, urgeatur corpus utrumq; secundum lineas parallelas, perget corpus primum describere circa corpus alterum areas easdem ac prius: vis autem qua corpus alterum urgebatur, jam destruetur per vim sibi æqualem & contrariam, & propterea (per Leg. 1.) corpus illud alterum vel quiescet vel movebitur uniformiter in directum, & corpus primum, urgente differentia virium, perget areas temporibus proportionales circa corpus alterum describere. Tendit igitur (per Theor. 2.) differentia virium ad corpus illud alterum ut centrum. *Q. E. D.*

Co-

A page from Newton's *Principia*

Newton's presentation copy to John Locke, showing corrections in his own handwriting

tion. The whole investigation is a perfect example of the true scientific method as described by Leonardo (p. 123). In the preface, Newton explains that the 'forces of gravitation with which bodies tend to the sun and the several planets' can be discovered 'from the celestial phenomena'. Having discovered what these forces are, he next deduces, by mathematical analysis, 'the motions of the planets, the comets, the moon and the sea' (Newton is here thinking of the tides). He continues: 'I wish we could derive the rest of the phenomena of nature by the same kind of reasoning from mechanical principles; for I am induced by many reasons to suspect that they all depend upon certain forces by which the particles of bodies, by causes hitherto unknown, are either mutually impelled towards each other and cohere in regular figures, or are repelled and recede from each other; which forces being unknown, philosophers have hitherto attempted the search of nature in vain.'

In accord with this plan, the first book inquires how the motion of bodies acted upon by known forces can be investigated mathematically. Galileo's experiments had disclosed the relation of motion to force, and Newton takes Galileo's mechanical system over complete, expressing it in the first two of his three 'Axioms or Laws of Motion'.

Law I. Every body perseveres in its state of rest or of uniform motion in a straight line, unless it is compelled to change that state by impressed forces.

Law II. Change of motion (i.e. rate of change of momentum) is proportional to the motive force impressed, and takes place in the direction in which such force is impressed.

Before enunciating these laws, Newton gave a number of definitions which were meant to explain the terms used in the laws.

These definitions and the 'Axioms or Laws of Motion' stand on much the same footing as the axioms of Euclid's geometry, and have met with much the same fate. Euclid thought that he could begin by stating a few axioms—truths which were so obvious as to need no proof—and deduce the whole of geometry from them. After they had stood unchallenged for some 400 years, Ptolemy noticed that they were not so much axioms as assumptions; now, after 2000 years, we have learned to regard them as specifications of the kind of space to which Euclid's theorems apply. In the same way, Newton's definitions and axioms stood more or less unchallenged for 200 years, until E. Mach, a Viennese professor, pointed out that the supposed definitions were not definitions but assumptions; they specified a particular kind of mechanical system— the particular kind of system to which the theorems of the *Principia* apply.*

Newton first defines mass, which, he says, is the volume of a body multiplied by its density. But as density can only be defined as the mass of a body divided by its volume, this takes us nowhere. What Newton's definition really does is tacitly to introduce the assumption that every body has associated with it a quantity possessing the various properties that he subsequently assigns to mass. One of these is that the mass of a moving object remains unchanged when the speed of the motion changes; we know now that this is not true (p. 295).

Newton next defines momentum ('quantity of motion') as being the mass of an object multiplied by its speed of motion, and this in turn raises the question as to how speed of motion is to be defined. We say that a car is going at 30 miles an hour when it passes a milestone every two minutes, but we must remember that the milestones are not themselves at rest, but are set in a rapidly moving and rapidly rotating earth. The car's motion along the road gives it a speed of 30 miles an hour past the milestones, but the milestones themselves have a further speed of perhaps 500 miles an hour caused by the west-east rotation of the earth's surface. On to this we must superpose a further speed of 70,000 miles an hour because the earth is moving in its orbit round the sun, and yet another

* *Die Mechanik in ihrer Entwickelung historisch-kritisch dargestellt*, 1883.

of perhaps 600,000 miles an hour because the sun is moving through the stars and so on, perhaps indefinitely. We know of nothing which is finally at rest and so can provide a fixed standard against which to measure motion. We may of course *assume* that if we went far enough, we should at last come to something possessing the needed property of absolute fixity, but this would be a pure hypothesis. Such, nevertheless, was the course that Newton took; he assumed that the remotest parts of the universe contained vast immovable masses, against which the motions of other objects could be measured. He saw clearly that it was only a working hypothesis: 'It may be that there is no body really at rest, to which the places and motions of other bodies can be referred.' And in any case he saw its difficulties: 'It is indeed extremely difficult to discover and distinguish effectively the true motion of particular bodies from the apparent; because the parts of that immovable space in which those motions are performed do by no means come under the observation of our senses.' We know now that the hypothesis of 'vast immovable masses' was unwarranted so that Newton's treatment of rest and motion was unjustified, and it was the same with his treatment of time. He says that time is something that 'flows equably without regard to anything external', but this is, of course, no definition; at the best it only records a single supposed property of time, and even this does not happen to be a true property.

Exactly 200 years after Newton had enunciated these hypotheses, an experiment by Michelson and Morley raised a suspicion that they were not universally true in nature, but the matter was not finally cleared up until Einstein's Theory of Relativity appeared early in the present century (p. 294). This showed that Newton's assumptions of absolute space and absolute time, as they are usually called, invalidate his mechanics for rapidly moving objects, including rays of light, although not for the more slowly moving objects to which Newton applied it.

The second law introduces the new conception of force,

but Newton is unable to define what force is. The furthest we get is that force is that which changes momentum, and that the amount of change is proportional to the amount of the force producing it. We have not yet been told how to measure force, but now the second law tells us: one force is twice as big as another if it produces twice as much change of momentum. Thus the second law is best taken as a definition of 'amount of force'.

Newton now introduces a third law:

Law III. To every action there is always opposed an equal reaction.

Thus the two forces which two objects exert on one another are equal in amount, but act in opposite directions. Here at last we are told something new—a real physical fact—about forces; if a horse draws a cart forward with a force x, then the cart exerts a backward drag x on the horse. If a gun fires a shot, and changes its momentum by y, then the gun acquires a recoil momentum equal to y. Laws I and II were taken over complete from Galileo, but Law III was Newton's own.

Galileo's mechanics had been limited to the earth and to the behaviour of terrestrial bodies, but Newton wished to show that the motions of the heavenly bodies could be explained on similar principles. He opens the third book of the *Principia* by saying that 'it remains that, from the same principles, I now demonstrate the frame of the system of the world', and proceeds to construct a complete mechanics of the heavens. He shows that if gravitation follows the law of the inverse square, then the planets must move exactly in accordance with Kepler's three laws. He discusses many other kinds of motion—as, for instance, the motion of a planet when other forces act on it besides the gravitational attraction of the sun. He also considers the motion which will ensue when three or more bodies mutually attract one another gravitationally, as, for instance, the sun, moon and the earth. This is the famous problem of the three bodies, to which no complete solution has yet been obtained.

Already in 1672 Richer had discovered that pendulum clocks were inclined to lose time in Cayenne and the same was found to be true of other regions near the equator. This showed that the earth's gravitation was less intense near the equator than elsewhere, and suggested that the earth might not be perfectly spherical, but orange-shaped—bulging at the equator and flattened near the poles—as Jupiter is seen to be. Newton showed that if the earth was flattened in this way, the pull of the moon on its equatorial bulge would tend to turn its equator into the plane in which the moon describes its orbit round the earth. This at once explained the 'precession of the equinoxes' which Hipparchus had discovered in the second century B.C.; the earth could not spin about a fixed axis as a steady top does, because the moon was forever turning the axis about.

Newton also showed how the gravitational pull of the moon and sun would produce tides in the oceans covering the earth, and discussed the theory of such tides in some detail, showing convincingly that the observed tides were just such as gravitation would produce.

Kepler had stated that all the known planets moved in ellipses round the sun, and Newton had showed that they must if their motion was determined by the sun's gravitation, yet they all moved in ellipses which were hardly distinguishable from circles. Newton's theory permitted bodies to move round the sun in ellipses of any degree of elongation, and he showed that some of the comets which had been observed in the past must have moved in very elongated ellipses. This particular part of the *Principia* assumed a special prominence after Halley had taken up the study of comets in 1680. He found that a comet which appeared in the sky in 1682 followed practically the same course as earlier comets which had appeared in 1531 and 1607. This suggested that all three might be appearances of a single body which was moving round the sun in a very elongated orbit which took $75\frac{1}{2}$ years to complete. Halley predicted that the comet would return again after another $75\frac{1}{2}$ years. It did so, and these strange objects, which had hitherto been regarded as portents of evil or of divine displeasure, were now seen to be mere chunks of inert matter which moved as the law of gravity compelled them to move.

Still another important section of the *Principia* dealt with motion in fluids, and acquired a special importance from its bearing on Descartes's theory of vortices. Newton showed that the planetary motions which were produced in the way that Descartes had imagined could not conform to Kepler's laws. This struck a mortal blow at the theory of vortices. Nevertheless, the theory lingered on for a time on the Continent, and such prominent scientists as the French astronomer de Fontenelle (1657–1757) and the mathematical physicist Johannes Bernoulli of Groningen and Basle (1667–1748) remained disciples of Descartes to their last days. It was particularly favoured in France, until Voltaire's advocacy of Newtonian theories gave it its long-deserved rest here also.

Later Telescopic Astronomy

Little remains to be said about the astronomy of the seventeenth century. In dynamical astronomy, Newton had advanced so rapidly, and left his fellow-workers so far behind, that it was long before more progress was made along the road he had so brilliantly opened. In observational astronomy, the telescope had opened up immense possibilities of which observers were not slow to take advantage. We have seen how Galileo had noticed two apparent appendages to Saturn in 1610 which he interpreted as small spherical bodies touching the larger spherical body of the planet at the opposite end of a diameter (p. 173). Hevelius found that these came and went with a regular periodicity, but their true nature was unknown until 1655, when Huygens saw that they were the protruding parts of a thin ring which surrounded the planet at its equator, and was so thin as to become invisible when its plane passed through the earth. In the same year he discovered Saturn's principal satellite, Titan. Within the next 30 years, Cassini, the leading spirit of the Paris Observatory, discovered four more. He also discovered that the ring which Huygens had seen round Saturn is divided into two separate parts by what we still call the 'Cassini' division. After this, discoveries in descriptive astronomy came thick and fast.

The telescope of these early days was admirable for revealing the wonders of the heavens, but was less suited for use as an instrument of precision, since it supplied no means for making exact measures of the positions and motions of the heavenly bodies. The value of the telescope was immensely increased as soon as subsidiary devices were invented to measure distances and motions in the sky with accuracy. Foremost among these was the micrometer, an instrument which took many forms and was certainly invented more than once. The thread of a spider's web is exceedingly fine, but a strand of it, put in a suitable position in a telescope, is thick enough to blot out the light from a selected star, so that none of it will enter the eye of a person looking through the telescope. If it is moved to some other position, the same spider-thread may blot out the light from a second star. If, then, an observer can measure the distance through which the spider-web has been moved since it blotted out the first star, he can estimate the precise distance between the two stars in the sky—not, of course, the distance in miles, but the distance in angle, the angle through which the telescope must be turned for one star to occupy the position that was previously occupied by the other. An instrument on these lines was designed and used by the Englishman William Gascoigne in about 1640, but his invention seems to have died with him. Other micrometers, working on essentially the same principle, were subsequently devised by Huygens and by Auzout and Picard of the Paris Observatory, and these enabled astronomers to measure small angular distances with great accuracy.

The invention of the pendulum clock by Huygens provided the analogous means for the accurate measurement of small intervals of time.

Armed with these and other similar instruments, astronomers were free to proceed to an exact study of the phenomena of the sky, measuring and cataloguing the positions and changes of position of the heavenly bodies with an ever-increasing degree of accuracy. In 1672 Cassini in Paris and Richer in Cayenne collaborated in determining the distance of Mars, using precisely the same method as the ordinary surveyor uses to determine the distance of an inaccessible point, such as the top of Mount Everest. From this they were able to deduce the distance of the sun, or the radius of the earth's orbit, and the general dimensions of the solar system. They estimated the distance of the sun to be about 87,000,000 miles, which may be compared with the best modern determination of 93,003,000 miles. Similar measurements were made for other bodies of the solar system, but the distances of the stars were not measured until 1838.

The details of these advances need not delay us here, except for one which was important because of its far-reaching consequences. We have seen how one of the first fruits of Galileo's astronomical work was the discovery of the four principal satellites of Jupiter. The innermost of these revolved round the planet once every 42½ hours or so, and was eclipsed once in each revolution as it passed into the shadow of Jupiter. The other satellites showed similar but longer periodicities. If the revolutions of the satellites had repeated themselves at absolutely regular intervals of time, they would have provided a most valuable astronomical clock which, as Galileo had suggested, could be used to determine longitudes at sea (p. 177).

Cassini had calculated time-tables for the eclipses of Jupiter's satellites, which he thought could be used in this way, and were actually so used on land to determine the longitudes of various unexplored parts of the earth's surface.

But when the Danish astronomer Römer set to work in 1676 to reobserve the satellites with a view to improving the tables, he found that the eclipses were not observed at absolutely regular intervals; they were sometimes a bit in front of their time-table, and sometimes a bit behind. Next he noticed that they were in front when Jupiter was near to the earth, and behind when it was far away—the light by which he saw them arrived late after a long journey. This suggested that it had taken time to travel through space, and he found that all the observations could be explained on this simple supposition. The range of variation in the times at which the eclipses were seen, about 22 minutes, must be the time that light took to travel a distance equal to the diameter of the earth's orbit round the sun. Combining this with Cassini's determination of the size of this orbit, Römer concluded that light must travel at about 138,000 miles a second; its true speed is now known to be about 299,770 km. or 186,300 miles a second.

Two thousand years before Römer, Empedocles had taught that light requires time for its journey through space (p. 46). Aristotle had accepted this on the general ground that 'when a thing is moved, it is moved from one place to another, and hence a certain time must elapse during which it is being moved. There was a time when the ray was not seen, but was being transmitted through the medium.' Yet most moderns had taken the view that light travelled instantaneously through space. Kepler had believed this on the ground that light was immaterial, and so could offer no resistance to whatever forces were propelling it through space. Descartes believed it on the ground that we see the moon eclipsed at the very moment when it is opposite the sun, and not some time later, as would be the case if light took time to travel through space, although, as Huygens pointed out, this did not prove that light travels instantaneously, but only that its speed is too great to be detected by the rough measurements that can be made at an eclipse. Galileo too was unconvinced and tried to measure the time light took to travel to a distant mirror and back again, but without success; light travelled too fast for its time over terrestrial distances to permit of measurement by the methods available to Galileo. Römer, working with astronomical distances, had now proved beyond challenge that the travel was not instantaneous, but his result was not generally accepted until Bradley confirmed it by an entirely different method (p. 241).

PHYSICAL OPTICS

After astronomy, the science which made most progress in the seventeenth century was optics, with Newton again as the main contributor. We have seen how the early Greeks had known the fundamental laws of the travel of light—that it proceeded through empty space in straight lines, and was reflected by a mirror so that the incident and reflected rays made equal angles with the reflecting surface. Then Ptolemy, or possibly someone else of his time, studied refraction, and obtained a law which, although not exact, was good enough for his purpose.

With these laws to build on, optics was studied somewhat intensively by the later Greeks, by the Arabians—al-Hazen in particular (p. 107)—and by the medieval European scientists from Roger Bacon on (p. 116). They had learned how to make lenses and mirrors, and understood how these could be used to change the convergence or divergence of a beam of light, and bring its rays to a focus. This line of investigation reached its climax in the invention of the telescope early in the seventeenth century, and its improvement in the hands of Galileo, Kepler, Huygens and others. In 1621 Willebrord Snell, Professor of Mathematics in the University of Leiden, discovered the correct law of refraction, but did not publish it, so that it lay unknown until Descartes announced it in 1637, whether as the result of an independent discovery or otherwise is not known.

Out of this the science of optical instruments has arisen;

its basic principles had been settled by the end of the seventeenth century, and its later development in matters of detail and technology need not concern us here.

This all forms a part of what we now describe as 'geometrical optics', which deals with the purely geometrical problems of finding what path will be taken by rays of light travelling without deviation except when they undergo reflection or refraction at the interface of a new medium. There is a second branch of optics known as 'physical optics', which deals with the very different question of what light is, and of why it behaves as it does, and this had hardly come into existence by the beginning of the seventeenth century.

We have seen how Euclid and Ptolemy had taken a wrong view of the nature of light and vision. They had followed the Pythagoreans in regarding light as an emanation from the eye, which gropes about until it falls upon the object for which the eye is searching. But al-Hazen had given a correct explanation of the act of vision in a treatise which was still the standard text-book on optics in the seventeenth century.

Kepler developed al-Hazen's doctrines further in his *Ad Vitellionem Paralipomena* (1604) and *Dioptrice* (1611). He described sight as the 'sensation of a stimulation of the retina', and said that the crystalline lens of the eye forms an image of the object of vision upon the retina. He thought that the retina contained a subtle spirit, the 'spiritus visivus', which was decomposed when light fell on it through the crystalline lens, much as combustible matter undergoes chemical change when sunlight falls on it through a burning glass—a remarkable anticipation of the 'visual purple', chemical changes in which are now known to produce the sensation of seeing. He pointed out that the chemical change in the 'spiritus visivus' must be rather lasting, since a retinal image persists for a short time after looking at a bright light. He explained short-sightedness and long-sightedness, quite accurately, as resulting from the crystalline lens of the eye bringing rays of light to a focus which does not lie on the retina. He also explained, again quite accurately, that we estimate the distance of an object from the small difference between the directions in which our two eyes look to see it—we subconsciously 'solve' the triangle which has the distant object as its vertex and the line joining our two eyes as base.

This all represented substantial progress, but it did not convince everyone. In particular, Descartes reverted to the old idea that light is an emanation from the eye.* We see surrounding objects, he said, as a blind man feels them with his stick. He gave precision to these ideas by supposing that light was propagated instantaneously by pressures transmitted from particle to particle of the medium which he supposed to fill all space. Building on this false basis, Descartes constructed an erroneous and unconvincing proof of the fundamental laws of optics, which gained some credence at the time and undoubtedly delayed the general acceptance of Römer's discovery of the finite velocity of light. He gave some thought to the nature of light, and conjectured that differences of colour were produced by particles rotating at different speeds, the most rapidly rotating particles producing a sensation of red in the eye, while slower rotations produced sensations of yellow, green and blue in this order.

But no one had so far given a convincing answer to the fundamental question 'What is light?' or a satisfactory explanation of the origin and meaning of colour. Many conjectures were in the air, but there was not enough experimental knowledge available to test them out. With the second half of the century came a great accession of knowledge of just the kind needed.

GRIMALDI (1618–63). In 1665 a book, *Physico-Mathesis de Lumine, Coloribus, et Iride*, appeared at Bologna, the posthumous work of a Jesuit, Francesco Grimaldi, who had been Professor of Mathematics in the University. It contained the first description of the phenomenon of 'diffraction'—the

* In his two books, *Le Monde* written in or about 1630, which was not published until 1664, and *Principia Philosophiae*, published in 1644.

phenomenon in which the undulatory quality of light is shown most clearly and most convincingly—and gave an account of a number of experiments on the subject. It will be enough to consider one of the simplest of Grimaldi's experiments.

In fig. 28 two screens, in which small holes, *a*, *b*, have been cut, are seen in cross-section, while below them lies a third screen, parallel to both, which has been left intact. If a strong light is put directly above the opening in the top screen, part of the bottom screen will be illuminated, and part left in shade. If light travelled in straight lines, it is easily seen that every part within the range *c* would receive some illumination, weak or strong, but the parts outside this range would remain completely dark, since light could only reach them by bending

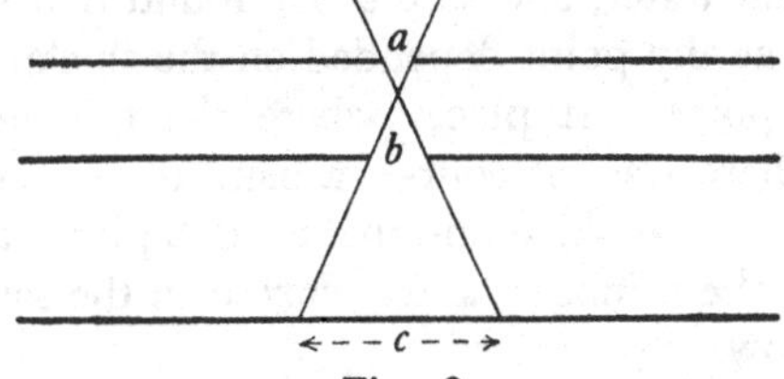

Fig. 28.

round a corner. But Grimaldi found that in actual fact the illumination extended well beyond the range *c*, thus showing that light was not limited to travelling in straight lines. Grimaldi and subsequent experimenters performed many variations of this simple experiment, and all gave the same result. Grimaldi further found that there was not a sharp transition from light to darkness at the edge of the shadow, but a rainbow-like band of colours, which he thought must have the same origin as the rainbow. The edge of the shadow was not only coloured but showed rhythmical alternations of light and darkness. These reminded him of the succession of ripples which appear when a stone is thrown into a pond, and led him to suppose, as Leonardo had supposed before him, that light is somehow associated with wave-motion. He obtained similar bands of colours by reflecting sunlight from a metal plate on which he had scratched a great number of close parallel lines. This roughly made instrument was the first example of the 'diffraction grating' which now forms an essential item in the equipment in every optical laboratory.

Here was a mass of experimental evidence bearing on the questions of the nature of light and the meaning of colour. But Grimaldi produced no useful suggestions beyond insisting that colour is a modification of light which somehow results from the detailed structure of matter.

HOOKE. In the same year 1665 another book, *Micrographia*, was published in England by Robert Hooke (p. 182). He had experimented with various thin layers of substances in which iridiscent colours could be seen—flakes of mica, soap bubbles, layers of oil floating on water, and so on. He found that the colour that was seen at any point depended on the thickness of the layer at the point. At places where the thickness changed gradually, there was, of course, a band of continuously varying colour, and Hooke found that at such places the colours were those of the rainbow, and occurred in the same order as in the rainbow.

Hooke imagined that ordinary white light was produced by rapid vibrations of particles in the luminous body, and that these vibrations spread out from the luminous body in the form of spherical pulses. Colour occurred whenever the symmetrical outward flow of these pulses was disturbed. The 'fundamental colours', blue and red, resulted from the pulses becoming 'oblique and confused' in different ways—blue when the weaker part of the pulse travelled first, and red when it came last.

NEWTON. Into the midst of all this confusion of thought came Newton, to whom, in Einstein's words, 'Nature was an open book whose letters he could read without effort. The conceptions which he used to reduce the material of experience to order seemed to flow spontaneously from experience itself, from the beautiful experiments which he ranged in order like playthings, and describes with an affectionate wealth of detail. In one person he combined the experimenter, the theorist, the mechanic and, not least, the artist in exposition. He stands before us, strong, certain and alone.'*

While still a student at Cambridge, Newton had read Kepler's *Dioptrice*, and had occupied himself in grinding lenses, and thinking about the performances of telescopes. In 1666 he bought a prism at Stourbridge Fair, near Cambridge, in order to 'try therewith the celebrated phenomena of colours'. His use of this phrase shows that the prismatic colours were well known at the time; indeed, we know that the diamond trade tried to cut their stones so as to show the colours at their best. A prism which was on sale at a local fair can have been little more than a toy, but with this toy Newton was to discover the secret of colour.

No discoveries seem to have come at first. For shortly after buying it he was helping Dr Barrow, his predecessor in the chair of mathematics at Cambridge, to revise his optical lectures for publication. They appeared in 1669, the year in which Newton succeeded to Barrow's professorship, and in them Barrow expresses quite fantastically impossible ideas as to the meaning of colour—white is a 'copious' light; red a condensed light interrupted by shady interstices; blue a rarefied light, as in bodies in which blue and white particles alternate, such as in the clear ether, or in the sea where white salt is mixed with black water; and so on. We can hardly imagine Newton passing all this for press if he had any better knowledge at the time.

His first account of experiments with his prism was published in 1672. It was his first scientific paper, and one of the very few he published without pressure and persuasion from his friends. For it was so fiercely attacked, and involved him in so much controversy that he seems to have become disgusted with all scientific discussion, and repugnant to the publication of anything at all from a fear of criticism and controversy; he took criticism badly.

* Foreword to Newton's *Opticks*, reprint of 1931.

His very first experiment revealed the true meaning of colour. He had made a small hole in the shutter of his room, through which a beam of sunlight could enter the room and pass through his prism (fig. 29). He found that the beam was spread out into a coloured band of light—a 'spectrum'—in which all the colours of the rainbow, from red to violet, were to be seen in the same order as in the rainbow; the band was about five times as long as it was broad. Anyone else might have done the same thing, and doubtless many had; the difference was that Newton set to work to wonder why the spectrum should be drawn out in this way. The presence of the prism had of course displaced both the violet

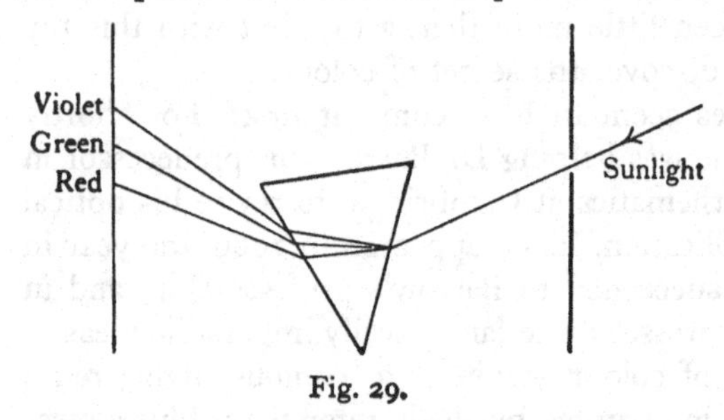

Fig. 29.

and the red light from the position that they would otherwise have occupied, but the drawing-out showed that it had displaced the violet farther than the red. The first part of the secret was now out—different colours meant different degrees of refrangibility; a ray of violet light was refracted through a greater angle than one of red light when it encountered a refracting surface. To test this conclusion, Newton made the various coloured lights in his spectrum undergo a second refraction in a direction at right angles to the first. He did this in order to see whether they would again merely experience the same degree of displacement or would undergo other and new changes, as, for instance, disintegrating still further into other colours. He found that the different colours of light retained their identities through the second refraction;

red remained red, and violet remained violet and each experienced precisely the same amount of refraction as before. He continued extending and checking his conclusions in a series of beautifully designed experiments, and finally announced his conclusion that sunlight was a mixture of lights of all the colours of the rainbow, the colours being permanent—or as Newton said 'original and connate'—qualities of the various ingredients, which experienced refraction in different degrees.

This gave at least a partial answer to the question 'What is colour?', but the more fundamental question 'What is light?' still remained unanswered.

Now although Newton's theories on colour had been tested and checked in every imaginable way, they had not carried conviction to everybody. Hooke and many others criticised them, and a lively and prolonged discussion resulted, in which the question of the nature of light frequently came into prominence. Newton hesitated to commit himself on this question, but he had to refer to other people's views in defending his own theory of colour. For instance, he suggested in 1675 that Hooke's hypothesis mentioned above (p. 204) could best be interpreted by supposing that bodies could be excited so as to cause vibrations in the ether 'of various depths or bignesses', and that the largest of these beget a sensation of a red colour, the least or shortest of a deep violet, and intermediate sizes sensations of intermediate colours, just as waves in air 'according to their bignesses, make several tones in sound'. In brief, he suggested associating different colours with ether vibrations of different wave-lengths, which is precisely what the undulatory theory did a century later. But Newton was propounding these interpretations of light and colour only as improvements on a theory of Hooke, and not as his own beliefs. Indeed he says at once that he likes another theory better, and this is a mixture of corpuscular and undulatory theories. Light begins as particles which excite waves in the ether, but are not themselves waves. 'Assuming the rays of light to be small bodies emitted every way from shining substances, these, when they impinge on any refracting or reflecting superficies, must as necessarily excite vibrations in the ether as stones do in water when thrown into it', and he suggests that many optical phenomena may be explained on the supposition that the various vibrations are 'of several depths or thicknesses, according as they are excited by the said corpuscular rays of various sizes and velocities'. He says that, if he had to assume a hypothesis it should be the foregoing but 'propounded more generally so as not to determine what light is, further than that it is some thing or other capable of exciting vibrations in the ether'. Nevertheless, he propounds a definite hypothesis to illustrate his ideas to those who need hypotheses.

He supposes that there is 'an aetherial medium, much of the same constitution as air, but far rarer, subtiler, and more strongly elastic'. This is not uniform, but consists of 'the main phlegmatic body of aether' mixed with various ethereal spirits; Newton considers that electric and magnetic phenomena and gravitation all seem to argue for such a mixture. As regards electrical phenomena, he thinks that when a piece of electrified glass attracts small bits of paper, the motion of the paper must be produced by 'some kind of subtile matter condensed in the glass, and rarefied by rubbing'; this rarefied ether may circulate through the surrounding space, carrying the papers with it, and finally return to the glass and recondense there....And as regards gravitation, he thinks that the gravitational attraction of the earth may be caused by 'the continual condensation of some other such like aethereal spirit, not of the main body of phlegmatic aether, but of something very thinly and subtilely diffused through it, perhaps of an unctuous or gummy tenascious and springy nature'. Yet light 'is neither ether, nor its vibrating motion, but something of a different kind propagated from lucid bodies'. This may be imagined either as 'an aggregate of various peripatetic qualities' or as 'multitudes of unimaginable small and swift corpuscles of various sizes' which spring from shining bodies and continually increase their speed until the resistance of the ether checks them—much as bodies falling in water are accelerated until the resistance of the water equals the force of gravity.

This is all very vague, and yet it is the only hypothesis as to the nature of light to which Newton seems ever to have given a modified approval. A strange mixture of the corpuscular and undulatory theories of light, it obviously aimed at securing the advantage of both, but it was meant only to aid the imagination, and not as a statement of absolute truth.

On this, as on many other questions on which Newton had no clear-cut views of his own, he put forward suggestions in the form of 'Queries', which he published at the end of his *Opticks*. In query 18, having asked whether radiant heat may not be transported by a 'much subtiler medium than air', he continues: 'And is not this medium the same with that medium by which light is refracted and reflected, and by whose vibrations light communicates heat to bodies?' But in query 29 he asks: 'Are not the rays of light very small bodies emitted from shining substances? For such bodies will pass through uniform mediums in right lines without bending into the shadow, which is the nature of the rays of light.' And finally (query 30): 'Are not gross bodies and light convertible into one another, and may not bodies receive much of their activity from the particles of light which enter their composition?'

Such seem to have been Newton's final thoughts as to the nature of light, all put forward very conjecturally. Throughout the main body of the *Opticks* he had been careful to insist that his results did not depend on any special views as to the nature of light, and had obviously tried to avoid the use of words which might seem to imply special views. The plain fact seems to have been that he was never able to make up his mind as to whether light was corpuscular or undulatory; he usually wrote as though it began as corpuscles and ended as vibrations which the corpuscles had excited in an ether. But as corpuscles were easier to understand than undulations, and provided a more obvious explanation of the linear propagation of light, the idea got spread about that Newton had declared light to be corpuscular.

The science of 250 years later again went through a phase of being unable to decide whether light consisted of waves or of corpuscles. For a time it was even thought to consist of both, but we now know that it consists of neither. At this time it was frequently said that Newton had shown great insight and brilliant prescience in advocating a corpuscular theory, and rejecting an undulatory theory. But Newton never did this, and even if he had, the praise would have been quite unwarranted. Newton's aim was to find a theory which would explain all the facts which were known about light in his time. Of these facts not a single one, as we now know, required a corpuscular theory; every one could, without exception, be explained on an undulatory theory, and was subsequently so explained. The facts which showed a pure undulatory theory to be inadequate did not become known until the end of the nineteenth century. Thus Newton's supposed repudiation of the undulatory theory and his advocacy of a corpuscular theory must have hindered the progress of optics very seriously.

HUYGENS (1629–95). While Newton was still thinking out his views on the nature of light, and before he published his *Opticks*, another theory was under construction in Holland by Christian Huygens, the son of a Dutch poet and diplomat. Huygens saw nothing corpuscular in light, and was content to regard it as wholly undulatory. His theory was first published in a paper that he read to the Paris Académie des Sciences in 1678, and was given in its full and final form in his *Traité de la Lumière* (1690). Like Descartes, Newton and others of the period, he imagined the whole of space to be filled with a medium, a 'very subtle and elastic medium', and supposed that a luminous object set up disturbances in this medium at perfectly regular intervals of time. These regular impulses produced regular undulations in the medium which were propagated in all directions in the form of spherical waves. Huygens supposed that each point on such a wave was occupied by a disturbance which itself formed a new source of spherical waves; in this way the waves propagated themselves.

On this supposition he was able to account for a number of the observed properties of light. The law of reflection of light was readily explained, and the additional assumption that light travelled less rapidly in dense media led at once to Snell's law of refraction. Huygens also gave a proof that light would travel in straight lines, but this was not altogether free from objections. And there were other phenomena before which the theory failed completely.

The Polarisation of Light. If a slab of glass is laid on a page of print, we see the letters clearly, although a little displaced on account of the refraction they experience in passing through the glass. But if a slab of calcite (Iceland spar) is laid on the page, we see each letter double, since calcite has the remarkable property of breaking any ray of light which passes through it into two distinct rays which travel in different directions—the property of 'double refraction', which the Danish physicist Erasmus Bartholinus discovered in 1670. Huygens's theory explained some of the properties of calcite very successfully, but failed completely with others. Newton's theories also failed, although he got nearer to the truth than Huygens had ever done, when he suggested that the phenomena needed rays of light to have 'sides'. He was here introducing the conception of what we now call the 'polarisation' of light. It permits of a very simple interpretation, but Newton missed it.

In a sound wave, each particle of air moves in what is called 'longitudinal' motion; it moves to and fro along the direction in which the waves are travelling. For instance, in waves of sound from a bell, each particle of air moves alternately towards and away from the bell. It is different with the waves of water on a pond; the waves travel along the surface of a pond, but the individual particles of water do not; their motion is up and down, and so at right angles to the surface; it is 'transverse' to the direction in which the waves are travelling.

Huygens's waves had been in longitudinal motion, but Hooke now suggested that light might consist of transverse waves, so that each particle of ether moved at right angles to the direction in which the light was travelling, and this implied that a ray of light had 'sides' in the Newtonian sense—two sides in the direction in which the particle moved, and two more in a direction which was at right angles to this and also to the direction of the wave.

If only these suggestions of Hooke and Newton could have been combined with the results obtained by Huygens, the theory of light might have been put on a satisfactory undulatory basis, for there were no facts then known which such a theory could not have explained. But, as things were, the corpuscular theory prevailed, supposedly backed by the high authority of Newton, and retained possession of the field until the end of the next century, when Young (1773–1829) and Fresnel (1788–1827) showed that all the facts then known could be reconciled in a pure undulatory theory (p. 253).

THE STRUCTURE OF MATTER

After optics, the most noteworthy advances in the physics of this period were those concerned with the general structure of matter, and with the interpretation of heat and fire. The atom was reinstated as the fundamental unit of which all matter was composed, and the chemical elements were discovered.

We have seen how Leucippus, Democritus and Epicurus had speculated as to an atomic structure of matter in the fifth century B.C. Their doctrines had been opposed at the time on account of their supposed anti-religious tendencies, and the

Middle Ages had neglected them entirely. They were now revived by a German botanist Joachim Jung (1587–1657), who lived and taught in Hamburg, by the French philosopher Pierre Gassendi (1592–1655) in his book *Syntagma Philosophicum* (1649), and by Robert Boyle (1627–91), who, it is said, was described as 'the Father of Chemistry and the Uncle of the Earl of Cork'.

GASSENDI. Jung was interested only in the botanical aspects of atomism, so that his speculations need not concern us further, but Gassendi made important contributions to physical thought. He imagined all matter to be composed of atoms which were absolutely rigid and indestructible; they were similar in substance, but varied in size and form, and moved about in all directions through empty space. He thought that many of the observed properties of matter could be accounted for by the motion of such atoms, as we now know they can. He also gave a tolerably accurate explanation of the three states of matter—solid, liquid and gaseous—as well as of the transitions from one state to another. But he went badly astray in conjecturing that heat in a body resulted from the presence of a special kind of 'heat atom'.

ROBERT BOYLE. Ten years later, Boyle became interested in these questions, and reached conclusions which helped to change chemistry from a mass of vague speculations into a coherent firmly knit science.

The scholastics of the Middle Ages had held the primitive Greek doctrine that all substances were blends of the four elements of earth, water, air and fire; while others had believed that the basis of every kind of substance was to be found in the three Arabian 'principles' of salt, sulphur and mercury. It was generally supposed that any substance could be 'resolved' into its fundamental elements or principles by fire. Boyle disproved this and many similar ideas by adducing quite simple facts against them.

He pointed out that some substances, such as gold and silver, withstand fire completely. When gold is subjected to fire, it does not yield earth, water or air, and neither does it yield salt, sulphur or mercury; it just remains gold. Even when gold is changed, as, for instance, when attacked by *aqua regia* (a mixture of hydrochloric and nitric acids), it still retains its existence and can be subsequently recovered as gold. It is the same when gold is mixed with other metals to form an alloy; there is no diminution or transmutation of its substance, and the alloy can always be made to yield back just the same amount of gold as was put into it. This, said Boyle, indicates that gold has a permanent and unalterable existence.

Other substances are, of course, altered by fire. In 1630 Jean Rey, a French physician, had proved* that tin and lead increased in weight when calcined. Boyle now repeated and confirmed his experiments, thus showing that calcination was something more than a mere resolution into more primitive substances.

Armed only with such simple considerations as these, Boyle challenged the existence of the four elements of earth, air, fire and water, and of the three principles of the Arabians. In 1661 he published a book,† in which the modern conception of a chemical element replaces the 'elements' and 'principles' which had obstructed the progress of chemistry for so long. Boyle explained that by elements he means 'certain primitive and simple bodies which, not being made of any other bodies or of one another, are the ingredients of which all those called perfectly mixed bodies are immediately compounded, and into which they are ultimately resolved'. A few years later, he stated‡ that all matter is made up of solid particles, each with its own determinate shape—the 'atoms' of modern

* *Essais de Jean Rey, docteur en médecine, sur la recerche de la cause pour laquelle l'Estain et le Plomb augmentent de poids quand on les calcine.*

† *The Sceptical Chymist: or Chymico-Physical doubts and Paradoxes, touching the Experiments whereby Vulgar Spagirists are wont to endeavour to Evince their Salt, Sulphur and Mercury to be the True Principles of Things.*

‡ *Origins of Forms and Qualities* (1666).

chemistry. Boyle says they can combine with one another to form the characteristic groups which we now call molecules. Such ideas might have put the study of chemistry on the right road, but they were somewhat slow at gaining general acceptance, and made but little progress until the French chemist Lavoisier advanced similar ideas a full century later (p. 263).

In the meantime Boyle was conducting a number of experiments on the physical properties of gases, with the collaboration of Robert Hooke, the very capable and ingenious experimenter whom he had secured as his assistant. Their work would have been impossible without the air-pump which Otto von Guericke, the Burgomaster of Magdeburg, had invented a few years earlier. Boyle and Hooke improved this instrument in 1659, and set to work to study the weight, compressibility and elasticity of air, as well as other of its properties. Boyle's first experiments with this 'new pneumatical engine' at once revealed what he called 'the spring of air'; when air was compressed it exerted force to regain its former volume, just as a coiled steel spring does. Exact measurements at different pressures led to the law, generally known in this country as 'Boyle's law' (1662), which states that if we double the pressure on a gas we halve its volume, and so on; the volume varies in inverse proportion to the pressure. The idea of such a law does not appear to have originated with Boyle, for he tells us that it had been suggested to him by Richard Townley, and that 'a certain person' (probably meaning Hooke), and also Lord Brouncker, the President of the Royal Society, had already made experiments which confirmed the law. In 1676 the French physicist Mariotte repeated some of Boyle's experiments and announced what was virtually the same law as Boyle had announced 14 years earlier; it is usually known as Mariotte's law on the continent.

Boyle and Hooke, still experimenting with the air pump, found that most combustible substances would not burn in a vessel from which the air had been extracted. With gunpowder, however, Boyle obtained curious results. When dropped on a heated plate in a vacuum it burned quietly but did not explode. He concluded that the nitre in gunpowder on heating gave 'agitated vapours which emulate air'. However, although Boyle had some idea of the part played by air in combustion, the first rational theory of the process was put forward by Hooke who stated that air was necessary for the 'dissolution of sulphureous bodies' and that this act of 'dissolution' produced 'a very great heat and that which we call fire'. He further stated that there was a constituent common to nitre and air. This was simply our oxygen but he did not succeed in isolating it.

In 1674 the Cornish physician John Mayow (1643–79) maintained that the increase of weight which a body experienced on combustion must result from its combining with some 'more active and subtle part of the air', which he called *spiritus nitro-aereus*; this was again oxygen, but again could not be studied in detail. Borch separated the same gas out from saltpetre in 1678, but still without understanding its nature, and it was not until Priestley rediscovered it in 1774 (p. 262) that its importance began to be realised.

MATHEMATICS

We have seen how mathematics lay almost stagnant throughout most of the Middle Ages. Geometry made but little progress after the Alexandrians had worked out the vein which had been started so brilliantly by the Ionian Greeks; new methods were needed if progress was to be resumed, and these were provided in due course by the analytical methods which we shall describe later (p. 219). Trigonometry was still regarded merely as a tool for astronomy, and had also lapsed into a state of stagnation. Algebra and arithmetic had made some progress, mostly in the sixteenth century, although this consisted mainly of improvements in notation and methods of calculation, the invention of logarithms provided the outstanding example.

In the seventeenth century, there was an advance along the whole front, accompanied by a change in the type of mathematics that was studied. In brief, applied mathematics came into existence, and pure mathematics fell back into second place. The Greeks had studied mathematics largely as a mental exercise—much as we work at a puzzle. They wanted more knowledge, because they found it intensely interesting in itself, and the process of acquiring and systematising it even more so, but they had little idea of developing the subject for its practical utility, and it was the same throughout the Middle Ages. A great change came in the seventeenth century when it was seen that the results of experiments or observation might need mathematics of a skilled kind for their adequate discussion.

The study of the conic sections provides an obvious example. Menaechmus and Apollonius had studied these purely for their intellectual interest, and without much thought of any practical application. The situation changed kaleidoscopically when Kepler discovered that the planets move in conic sections. The study of these curves assumed a new importance, and Newton's *Principia* carried it on to a still higher plane. Just at this time the methods of analytical geometry came into general use, and any tolerably competent mathematician could now solve by routine methods problems which would previously have called for all the skill of a great mathematician.*

Pure mathematics was still studied for its own sake, and for the fascination of its methods. There was a search for theorems which had no practical applications, and presumably no value beyond the satisfaction they gave to those who discovered them and those who contemplated them. A

* Newton must have been acquainted with these new methods, for he had read Descartes's *Geometry*, yet he did not use them in the *Principia*, but proved his theorems by the methods of the older geometry, possibly so that they might be more generally understood. He presents some of them in a form which suggests that they had been discovered by the new methods, and then translated back into the language of the old.

typical instance was the theorem of Fermat that no integral numbers x, y, z can be found such that $x^3+y^3=z^3$, and that the same is true of all higher powers than the cube—a theorem which Fermat stated without proof, and for which neither proof nor disproof has so far been found. Another example from a later period is the theorem that every even number can be expressed as the sum of two primes; although a vast amount of labour has been devoted to this, it has neither been proved nor disproved.

But these triumphs of abstract mathematics were mostly in the nature of side-shows, and the mathematicians gave their attention mainly to such investigations as were urgently needed for practical use, or gave promise of being of practical use, and herein the mathematics of the seventeenth century differed from that of all the preceding centuries. Its primary aim was the application of mathematics to the phenomena of nature; its main achievements were the development of analytical geometry, and the introduction of the infinitesimal calculus, which, in its two forms of the differential and the integral calculus, made mathematics specially suited to the investigation of natural phenomena.

Analytical Geometry

The discovery of analytical geometry is usually attributed to Descartes (1596–1650) and Fermat (1601–65), but it was certainly known and used before their time, and possibly even by Apollonius.

After Descartes had resigned his commission so as to have more time for mathematics and philosophy, he first wrote *Le Monde*, which he did not publish (p. 179), and then the most famous of all his books, *Discours de la Méthode*, which appeared in 1637. This was mainly philosophical, but Descartes added three scientific appendices to it in the subsequent year, the third being *La Géométrie*, in which he explained the principles of analytical geometry. This is the book which Newton read at Cambridge, and which helped to create his interest in mathematics. It is difficult to read, and its style is obscure. Descartes said he made it so on purpose, for fear that certain wiseacres might be tempted to say that they had known it all the time. Fortunately, John Wallis (1616–1703), a Cambridge mathematician who had become professor at Oxford, published a treatise on conic sections in 1655 which explained the whole subject very clearly.

Another great French mathematician, Pierre de Fermat was also interested in the subject, and seems to have discovered the general methods of analytical geometry independently of Descartes. But as most of his work remained unpublished until after his death, the credit must go mainly to Descartes.

The general principles of the method are readily understood. The various properties of the circle, as enumerated by Euclid, seem at first to be detached and independent properties—like the size of a man's shoes and the colour of his overcoat—but they are not. Most of them are true of no curve but the circle, so that any one of them forms a sort of definition, or complete description, of the circle, and so contains all the properties of the circle inherent in itself.

We may define the circle as a curve on which every point is at the same distance from a point we call the centre. This is Euclid's definition, and all the properties of the circle are inherent in it, as Euclid shows. But we might also define the circle as a curve such that a fixed line AB subtends the same angle at every point of it—i.e. as we walk round the circle, the line AB looks always the same size to us— the angle ACB in fig.31 remains always the same.* This seems at first to have but little relation to the Euclidean definition, yet it can be shown that the two definitions imply precisely the same properties; they are logically equivalent. Now the method of analytical geometry is to take a definition of yet another kind, which we may best describe as an algebraic definition, and

* There is a slight complication when we pass through either A or B where the angle subtended changes discontinuously.

deduce as many of the properties of the curve as we want from this definition by purely algebraic processes.

We may imagine a plane area mapped out into squares (fig. 30), much as a country can be mapped out by parallels of latitude and longitude, and we may specify the position of a point C in the area by saying that it lies x units of length to the right of a point O, and y units of length above a horizontal line through O. Or, if OX and OY are two perpendicular lines through O, we can say that the point C is at a distance

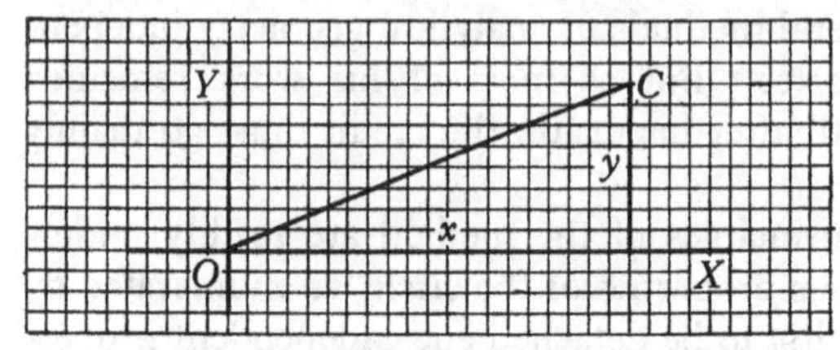

Fig. 30.

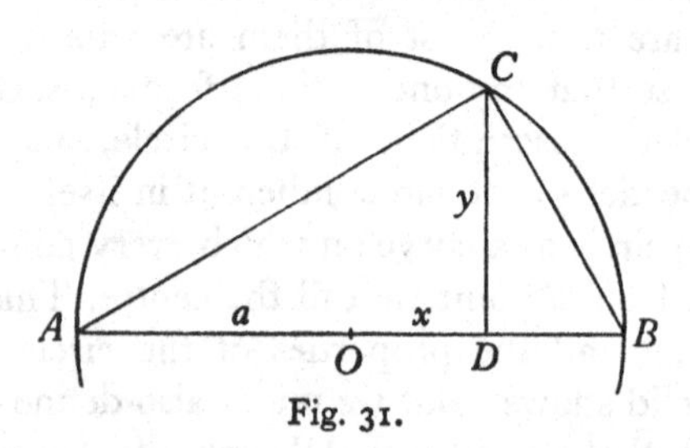

Fig. 31.

x along OX, and a distance y along OY. The values of x and y are called the 'coordinates' of the point C.

The theorem of Pythagoras now tells us that $OC^2 = x^2 + y^2$. Thus if a circle of radius a is drawn round O as centre, then C will lie on the circle if $x^2 + y^2 = a^2$ but not otherwise. Thus this relation is true for the coordinates of all points which are on the circle, and for those of no other points; it forms an algebraic specification of the circle, and all the properties of the circle are inherent in it. We call it the 'equation' of the circle.

We should have obtained precisely the same relation if we had defined the circle in any other way—for example, as a curve such that a fixed straight line AB subtends a right angle at every point of the curve (fig. 31). For let AB be of length $2a$, and let O be its middle point, and let a perpendicular from any point C on the circle on to AB meet it in D, where $OD=x$ and $DC=y$. Then ACB will be a right angle, and the theorem of Pythagoras tells us that

$$AB^2=AC^2+CB^2=(AD^2+DC^2)+(DC^2+DB^2)$$
$$=(a+x)^2+2y^2+(a-x)^2=2\,(a^2+x^2+y^2).$$

Since AB^2 is equal to $4a^2$, the equation becomes $x^2+y^2=a^2$, which is the same relation as we obtained before.

Whatever geometrical definition we take for the circle, we shall always arrive at this same algebraic description. Indeed we must, or the same curve would have two inconsistent sets of properties. And, reversing the procedure, just as this description can be derived from any one property of the circle so, conversely, any and every property of the circle can be derived from this description by simple algebraic processes.

In the same way, any other curve can have all its properties summed up in a single equation of a similar kind, this equation forming a sort of compendium of the properties of the curve, which may be many hundreds in number. We can deduce all the properties of the curve, or as many as we want, from this one equation.

Such are the methods of analytical geometry. They are incomparably more powerful, more concise and more penetrating than the groping methods of the old Greek geometry, so much so that the old geometry is now little more than a museum piece.

THE INFINITESIMAL CALCULUS

The other principal creation of the seventeenth-century mathematics was what we now call the calculus of infinitesimals. It may be described as a method of applying mathematics to continuous changes such as occur in nature, rather than to the simplified and unreal conditions such as are often imagined by mathematicians. The calculus takes many forms, but its inner essence can be illustrated by a very simple problem—to find the area enclosed by a curved line.

The area of a rectangle of height h and length l is of course known to be hl, the product of the height and the length. But we can also think of this area as the sum of a number of

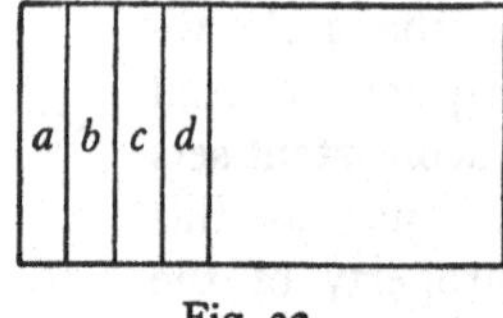

Fig. 32.

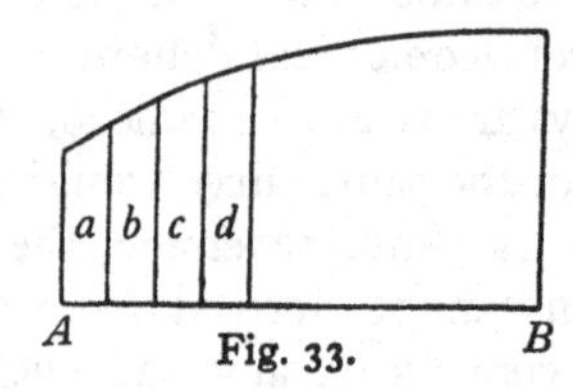

Fig. 33.

strips $a, b, c, d, \ldots$ as in fig. 32. If any strip has a width w, its area is hw, and the sum of all the hw's is the total of the area required.

This is very simple, but now suppose that the area is not bounded by a straight edge as in fig. 32, but by a curved edge as in fig. 33. The hw's are now all different, because the h's are different; as we pass along the area from A to B, h changes continuously. Let the new heights of the strips $a, b, c, \ldots$ be $h_a, h_b, h_c, \ldots$. Then the total area of which we are in search is $h_a w + h_b w + h_c w + h_d w + \ldots$.

There is, however, an ambiguity in the measure of the heights of the strips, and we shall obtain different results according as we take the height of the strip in fig. 34, to be AP or $A'P'$ or $A''P''$. Obviously the only way to get a consistent and exact result is to arrange that the heights AP, $A'P'$, $A''P''$ shall all be the same, and this we can do by making the strip of only infinitesimal width. The area we need is then the sum of an infinite number of strips each of which is zero area. This simple example contains in it the essence of the calculus of infinitesimals.

The method first appeared in the seventeenth century, when it was known as the 'method of indivisibles', the term indivisible meaning 'too small to be divided'. Kepler had used the method in a rather primitive form as far back as 1604, but it was first clearly stated by the Italian mathema-

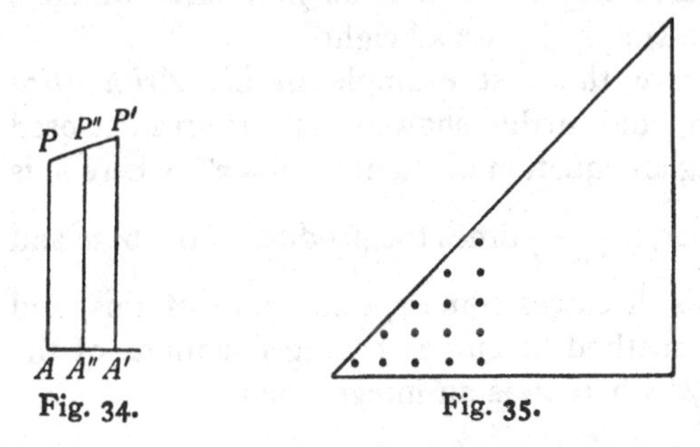

Fig. 34.

Fig. 35.

tician Bonaventura Cavalieri (1598–1647) in a book which he wrote in 1629 and published in 1635. Cavalieri thought of an area as being made up of a number of minute rectangles or 'points', all of which were equal and infinitesimal in size. A column of these points formed a strip like those we have already considered. If he wished to find the area of the triangle shown in fig. 35, he regarded it as the sum of strips which consisted of 1, 2, 3, ... points respectively. The total area of a triangle of n strips will be $1+2+3+\ldots+n$ points. Cavalieri knew that the sum of this series of numbers is $\frac{1}{2}n^2+\frac{1}{2}n$. The first term $\frac{1}{2}n^2$ in this sum is n times the second term, so that, as n is a very great number, it completely outweighs the second term, and the sum may, without error, be taken to be $\frac{1}{2}n^2$. Thus the area of the triangle is that of $\frac{1}{2}n^2$ points, which is equivalent to the ordinary formula, $\frac{1}{2}$ base $\times$ height.

In this special instance, the method leads to no new result, but in other cases it does. Suppose the area is not a triangle bounded by a straight line, as in fig. 36, but is a curved area bounded by a parabola, the equation of the parabola being $y=x^2$. Then the strips a, b, c, ... must be supposed to consist of 1^2, 2^2, 3^2, ... points, and the whole area is that of

$$1^2+2^2+3^2+\ldots+n^2$$

points. Now the sum of this series is $\frac{1}{3}n^3+\frac{1}{2}n^2+\frac{1}{6}n$, in which the first term is infinitely greater than those which follow. Thus the whole area may be taken to be $\frac{1}{3}n^3$ points, which is readily seen to be simply $\frac{1}{3}$ base $\times$ height.

John Wallis gave this last example in his *Arithmetica Infinitorum* (1656), and further showed that the area enclosed by a curve having an equation of the form $y=x^m$, where m is an integral number, is $\dfrac{1}{m+1}$ times the product of the base and the height. Wallis discusses some special cases of this, and then extends the method to curves having equations of the forms $y=(1-x^2)^m$ where m is an integer, and

$$y=a+bx+cx^2+ex^3+\ldots.$$

In the latter case, the area is found to be

$$x(a+\tfrac{1}{2}bx+\tfrac{1}{3}cx^2+\tfrac{1}{4}ex^4+\ldots).*$$

When Wallis tried to extend his results to more complicated problems, he met with a check. He would have liked to find the area of the circle of unit radius, of which the equation is $y=\sqrt{(1-x^2)}$ or $y=(1-x^2)^{\frac{1}{2}}$, but did not know how to 'expand' this into a series of the standard form $y=a+bx+\ldots$.

* The mathematician will see that the germs of the integral calculus had already emerged; Wallis had found the values of the quantities we now denote by $\int x^m dx$, $\int (1-x^2)^m dx$, and $\int (a+bx+cx^2+cx^3+\ldots)\,dx$. Shortly after this, Pascal gave the correct values of the quantities we now denote by $\int \sin\phi\, d\phi$, $\int \sin^2\phi\, d\phi$ and $\int \phi \sin\phi\, d\phi$.

From this point on, most of the progress was made in the first instance by Newton, although he was so dilatory in publishing his results that others inevitably got a large share of the credit. When he read Wallis's book in the winter of 1664–5, he saw the importance of discovering the expansion of $(1-x^2)^{\frac{1}{2}}$. He set to work and not only discovered this, but also the more general expansion of $(1-x^2)^m$, where m is any quantity, fractional or integer, positive or negative—the expansion that is usually known as the 'binomial theorem'. We have seen (p. 184) how Newton tells us that he obtained this expansion in the two plague years of 1665 and 1666, 'when he was in the prime of his age for invention', and 'minded mathematics and philosophy more than at any time since'. Yet he did not publish it until 1704, when it appeared in an appendix to the *Opticks*, entitled *De Quadratura Curvarum*. Thus was the binomial theorem first given to the world—nearly 40 years after its discovery.

The second part of this belated appendix was even more important than the first, for it explained the most useful of all Newton's mathematical discoveries, the differential calculus, which he called the 'method of fluxions'. He had started working on this in a manuscript which he had dated 1665.

Newton imagined that a quantity x, which he called the 'fluent', the flowing (or changing) quantity, changed its value continually with the progress of time. He described the rate at which it changed as the 'fluxion', and denoted this by $\dot{x}$. He wished to determine the relation between the fluent and its fluxion in precise mathematical terms.

Suppose, for instance, that the fluent x changes in such a way that after any number t of seconds have elapsed, its value is always at^2, whatever the number t may be—the law which Galileo found for a sphere rolling down an inclined plane. Newton first divides the time t during which change takes place into an infinite number of moments, each of infinitesimal duration o, and imagines that after a time t

another moment of duration o elapses. In this interval x increases by $o\dot{x}$, and so changes to $at^2+o\dot{x}$. But as a total time $t+o$ has now elapsed, the new value of x must be $a(t+o)^2$, or

$$at^2+2ato+ao^2.$$

Since this must be the same thing as $at^2+o\dot{x}$, the value of $\dot{x}$ must be $2at+ao$. The last term ao is infinitesimal in amount, and so may be 'blotted out', and we are left with $2at$ as the fluxion of at^2. If a body falls through a distance at^2 in t seconds, then its rate of fall must be $2at$.

In another type of problem, the fluxion is given, and it is desired to find the fluent. For instance, it is found, as a matter of observation, that the speed of a body falling freely through space increases by 32 feet per second in each second of the fall. After t seconds, the speed of fall will be $32t$ feet per second. Thus if the body has fallen through y feet, $32t$ must be the fluxion of y. We have just seen that $2at$ is the fluxion of at^2, so that $32t$ must be the fluxion of $16t^2$, which means that after a body has fallen for t seconds, it will have fallen through a distance of $16t^2$ feet.

Suppose that this same object was projected horizontally with a speed v just as its fall began. If we neglect air resistance, this horizontal speed will persist unaltered through the whole fall, so that in t seconds of fall the projectile will travel a horizontal distance x equal to vt. Thus we may replace t by x/v, and see that the value of y is $y=16x^2/v^2$. This relation between x and y is, of course, the equation of the path of the projectile; it represents a parabola. Here we see how the calculus can be used to investigate the motion of a body, while analytical geometry is used to interpret the results.

These simple illustrations will have given some slight idea of how these two new branches of mathematics opened all the phenomena of nature to exact study. No doubt the various problems could often be solved by other methods—actually those just given were first solved in other ways by Galileo. But other solutions depended on the exercise of ingenuity, and perhaps even on the smile of good fortune, in hitting upon the right way, while the method of fluxions reduced every problem to a matter of routine. Not every problem admits of solution in this way, but it is usually easy to see whether a problem is soluble and, if it is, to discover its solution.

If Newton had published his discovery immediately he made it, the value to science would have been very great. Actually he wrote out a sketch in 1669, which he passed on to Barrow, and probably also to a number of friends and pupils, but there was no publication in the ordinary sense. In June and October 1676—10 years after his invention of the method—we find him writing two letters to his contemporary Leibnitz, explaining what he had done. He was still unable to refer him to anything published, and concealed his achievement in the unintelligible cipher

6a cc d æ 13e ff 7i 3l 9n 4o 4q rr 4s 9t 12v x,*

after which he did nothing until 1704, when the substance of the matter was published in the appendix to the *Opticks*. A full account of the method first appeared in 1711.

At this time the progress of science was crying out for just some such method as Newton had discovered, while the work of Cavalieri, Wallis and many others was leading in the direction of just such a discovery. Thus Newton ran a grave risk of someone duplicating his discovery, and publishing it first. And someone actually did, in the person of Leibnitz.

LEIBNITZ (1646–1716). Gottfried Wilhelm Leibnitz was born and educated in Leipzig, studying mathematics, philosophy, theology and law before attaining the age of 20. After serving the Elector of Mainz for a time, he entered the diplomatic service. He next entered the service of the Brunswick family and in 1676 became librarian to the Duke of Hanover, a position in which his duties left him ample leisure for his favourite pursuits of philosophy and mathematics. In 1682,

* 'Date equatione quotcunque fluentes quantitates involvente, fluxiones invenire et vice-versa.'

conjointly with Otto Mencke, he founded a journal called *Acta Eruditorum*, the only privately owned scientific journal in Europe at the time.

Beginning in the issue of this journal for October 1684, he published a series of papers in which he developed an infinitesimal calculus which was substantially identical with that of Newton, but was expressed in a simpler and far more convenient form. His notation was that which we still use to-day, and was substantially superior to Newton's. Where Newton would have written $o\dot{x}$ for the increment in x, Leibnitz wrote $(dx/dt)\,dt$, or, more simply still, dx. Leibnitz shortened both the thought and the writing by discarding the idea that the changes must happen in time. And as Leibnitz began publishing his development of the subject in 1684, whereas Newton did not get started until 1704, Leibnitz got most of the credit in the eyes of the world, and from one point of view rightly, since the calculus would have been just as useful to the world at large if Newton had never discovered it. The situation, which must have been very galling to Newton, led to a long and bitter controversy, first between Newton and Leibnitz, and then—up to, and even after, their deaths—between their partisans. Neither side disputed that the methods of Leibnitz and Newton were substantially the same; neither side had disputed that Newton had been the first to discover the method, or that Leibnitz had been the first to publish it; the only point in dispute was whether Leibnitz, in spite of his many protestations to the contrary, had or had not got some or all of his ideas from seeing certain manuscripts of Newton. This he might conceivably have done, for while he was in London, a manuscript by Newton was lying at the Royal Society, and might have been shown to Leibniz. Fortunately, the squabble had no appreciable influence on the growth of science, and so forms no concern of the present book.

Leibnitz wrote many other mathematical papers, mostly in the *Acta Eruditorum*, but, outside his papers on the infinitesimal calculus, none was of first-rate importance. Many of them contained bad blunders, but at least they familiarised the mathematicians of Europe with the methods of the infinitesimal calculus. The same effect was produced by the writings of two brothers Bernoulli—James Bernoulli (1654–1705), who was born, lived and died in Basle, being Professor of Mathematics in the university there, and his brother Johannes (1667–1748) who succeeded to the chair on his death.

CHAPTER VII

THE TWO CENTURIES AFTER NEWTON
(1701–1897)

We now approach an era which, if not so strikingly brilliant as its great predecessor, was at least one of solid and steady progress. It produced no second Newton, but provided an abundance of first-class investigators. The studious and talented amateur could still accomplish scientific work of the highest value, for a single mind could carry a good working knowledge of a substantial part of science; the days of immense stacks of literature and teams of experts, each understanding only one corner of a subject of research, had not yet arrived. They were, however, on the way, for there was already a tendency for the various sciences to unite into one, to lose their identity as detached units, and become merged into one single field of knowledge which was too vast for anyone to comprehend the whole, or even a large fraction of it. We note the appearance of such words as thermodynamics, astrophysics and electrochemistry.

MECHANICS

Mechanics looms large in the story of eighteenth-century progress. Galileo and Newton had opened the road, but a lot remained to be done in extending their gains and filling up lacunae. Newton's laws of motion were applicable only to particles, i.e. to pieces of matter which were small enough to be treated as points, and so could have definite positions, velocities and accelerations unambiguously assigned to them. Every body is of course made up of particles, so that it is possible in principle to deduce its motion from Newton's laws, but the passage from the general principles to the solution of a particular problem may be long and difficult. There was an obvious need to find general methods for effecting this passage, and the problem attracted the attention of some of the greatest mathematicians of the age.

The problem assumes its simplest form when the body in question is 'rigid', i.e. when the distance between any two of its particles is unalterable, so that the body cannot change its shape. We have seen how the values of two coordinates will fix the position of a point in a plane (p. 220). In the same way, the values of three coordinates will fix the position of a point in space, and so will fix the position of any assigned point of a rigid body (e.g. its centre, if it has one). Three more quantities are needed to fix the orientation of the body in space, and these will generally be angles, or of the general nature of angles. Knowing the values of these six quantities, we can deduce the position of every particle of the rigid body, and if the changes of these six quantities can be traced out, we can follow the whole motion of the body. These six quantities may be described as the 'generalised coordinates' of the body, and rules for following their changes were determined by Euler.

LEONARD EULER (1707–83) was one of the great mathematicians of this period. The son of a Lutheran minister in Basle, he lived in succession in Basle, St Petersburg and Berlin, to which he moved on the invitation of Frederick the Great, and subsequently returned to St Petersburg, where he became totally blind and died in 1783. From Newton's laws for the motion of a particle, he deduced general laws for the motion of a rigid body, and these give satisfactory explanations of the movements of gyroscopes, spinning-tops, of the flight of a spinning golf-ball, of the precession (p. 92) and nutation (p. 241) of the earth, and a variety of similar motions.

LAGRANGE (1736–1813). Still greater progress was made by another protégé of Frederick, Joseph Louis Lagrange, who was probably the greatest of all the great mathematicians of this period. He was a native of Turin, but moved to Berlin at the age of 30, Frederick having expressed a desire that 'the greatest mathematician in Europe should reside at the Court of the greatest King in Europe'. When Frederick died in 1786, Lagrange received invitations from Spain, Naples and France to migrate to their capitals. Accepting the last of these, he lived in great style at the Louvre, although suffering continuously from poor health and fits of profound melancholy. Then the Revolution came and altered his status considerably. However, he stayed on in Paris as a professor, first at the École Normale and then at the École Polytechnique, until his death. He was chairman of the Commission which the French Government appointed in 1799 to reform its system of weights and measures, and it was largely through his efforts that the decimal system of measures, with the metre and gramme as units, was established in France, from which of course it rapidly spread over the continent of Europe.

Euler had shown how to transform Newton's laws so that they should apply to the motion of any rigid body. Lagrange now transformed them so that they were applicable to the most general system of bodies imaginable. Here, of course, the changes in six generalised coordinates are not enough to tell us the whole story of the system. But, however complicated a system may be, its configuration can always be described by a sufficient number of generalised coordinates, and the changes in these coordinates will tell us all about its motion. Lagrange showed how to obtain this knowledge by purely routine methods.

This was a great advance, but he continued it still further in a direction which we must now describe in detail.

Least Action. Hero of Alexandria had shown (p. 81) that a ray of light always followed a path of minimum length, and this is true even if the ray is reflected by one or more mirrors. As light always travels at the same speed in air, a path of minimum length is also a path of minimum time.

After Snell (p. 200) had discovered the true law of refraction of light, Fermat (p. 219) showed that refracted light conformed to this same principle of minimum time, provided that the speed of light depended on the substance through which it was travelling, in a way which is now known to be the true way. Assuming this, it can be shown that the total time of any journey is again a minimum. We can understand the bending of a ray of light by refraction if we think of the ray as showing the direction of a line moving at right angles to itself, like troops in line abreast or waves coming in to break on the sea shore. Fermat's assumption (which has since been verified by experiment) was that light travels more slowly in a dense medium, such as glass, than air; therefore the first part of the line to reach the refracting surface is slowed up and the whole line wheels round and takes a new direction. Thus

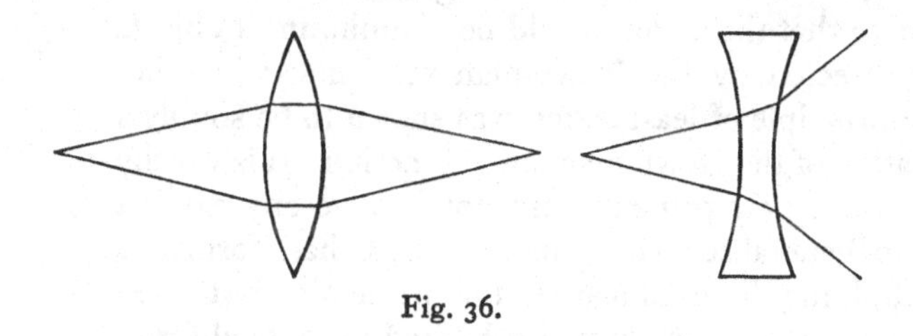

Fig. 36.

in fig. 36, a convex lens causes the rays to bend as shown on the left and a concave lens bends them as shown on the right. In this way a convex lens causes a beam of light to converge while a concave lens makes it diverge.

A century after Fermat, P. L. M. de Maupertuis (1698–1759) conjectured that all natural movements must conform to some similar principle.* His reasons were theological and metaphysical rather than scientific; he thought that the perfection of the universe demanded a certain economy in nature which would be opposed to any needless expenditure of activity, so that natural motions must be such as to make some quantity a minimum; it was the kind of general principle that would have made a great appeal to the Greek mind —and it was successful.

* *Essai de Cosmologie*, 1751, p. 70.

The main difficulty was to find the quantity in question. It could no longer be the time, since to make this a minimum all objects would have to dash through space at the highest speeds of which they were capable—and this was clearly not the way of nature. Maupertuis introduced a quantity which he called the 'action' of a motion. It was the time of the movement multiplied by the average value of the *vis viva* throughout this time,* and he thought that this quantity ought to assume its minimum value when bodies moved in their natural way; he called this the principle of least action. Euler had been favourably impressed by the idea, and had advanced arguments in its support. Lagrange now produced proof positive that the action would be a minimum if objects moved as directed by the Newtonian mechanics; in other words, the principle of least action was shown to be simply a transformation of the Newtonian laws of motion. It is worthy of note that the new principle did not involve any explicit reference to generalised co-ordinates; these had formed a useful scaffolding in establishing the principle, but were removed before the principle was exhibited in its final form.

This principle reduced every problem of dynamics to a problem of algebra, just as analytical geometry had reduced every problem of geometry to one of algebra. Lagrange published his proof of the principle in his *Mécanique Analytique* (1788), and wrote in the preface: 'We have already several treatises on mechanics, but the plan of this one is entirely new. I intend to reduce this science, and the art of solving its problems, to general formulae, the simple development of which provides all the equations needed for the solution of any problem....The methods that I explain require neither geometrical nor mechanical construction or reasoning but only algebraic operations.'

In this way the laws of propagation of light and the laws

* *Vis viva* meant twice the quantity that we now call the kinetic energy, i.e. the quantity obtained on multiplying the mass of each moving body by half the square of its velocity, and adding up the contributions from them all. In mathematical language, *vis viva* is Σmv^2.

of the motion of material bodies were shown to be similar in form; in each case a certain quantity assumed its minimum value. This similarity in form explains why some mechanical processes remind us of other optical processes—for instance, the rebound of a tennis ball reminds us of the reflection of a ray of light, and the curved path of a projectile moving under the earth's gravitation is like the curved path of a ray of light in the earth's atmosphere. The analogy was not altogether helpful to science, since it ultimately led the nineteenth-century physicists to imagine that the propagation of light could be explained in terms of Newton's laws, and this we now know is not so.

HAMILTON. In 1834, Lagrange's formulae were further transformed by Sir William Rowan Hamilton (1805–65), the son of Scottish parents living in Dublin, who was elected Professor of Astronomy at Trinity College, Dublin, at the age of 22, while still an undergraduate. He put Lagrange's formulae into the form known as 'canonical equations', in which they closely resemble Newton's original equations, but with the all-important difference that they do not deal with the motions of infinitesimal particles but with the changes in generalised coordinates. Newton's laws had said that the rate of change of momentum (of a particle) was equal to the force; Hamilton's laws now said that the rate of change of a quantity called the generalised momentum is equal to another quantity called the generalised force. In the simplest case of a particle acted on by forces, the Hamiltonian laws reduce exactly to the Newtonian laws; in general, the Hamiltonian laws are adequate to deal with the most complicated system of bodies imaginable.

Dynamical Astronomy

The first science to benefit from all this was rather naturally astronomy, and the principal benefactor was Lagrange's great contemporary Laplace, who has been described as the French Newton.

LAPLACE (1749–1827). Pierre Simon Laplace was born in Normandy on 23 March 1749, the son of a cottager. Some well-to-do neighbours paid for his education at a school in which he afterwards became an usher. Later he went to Paris, and wrote a letter on the principles of mechanics to D'Alembert, who was so impressed by it that he obtained for Laplace a position in the École Militaire at the early age of 20. Sixteen years later (1785) it fell to Laplace to examine the young cadet Napoleon Bonaparte in mathematics.

His appointment in the École Militaire set Laplace free to engage in original scientific research, and the next few years saw an immense output of important papers on a variety of mathematical, physical and dynamical subjects. He was not, however, satisfied to shine only in scientific circles; he wished to cut a figure in social and political life as well. The French Revolution gave him the opportunity he craved for, and he became a hanger-on, first of the mob and then of Napoleon. His active political career began when he persuaded his former examinee to appoint him to office as Minister of the Interior. It ended six weeks later. Rouse Ball* has transcribed the minute with which Napoleon recorded his dismissal: 'A geometer of first rank, Laplace lost no time in showing that as an administrator he was worse than mediocre; his first piece of work showed that we had been deceived in him. He could see nothing from a realistic point of view; he looked for subtleties everywhere, had only problematical ideas, and carried the spirit of infinitesimals into administration.' Nevertheless, Napoleon elevated him to the Senate, probably from a desire to remain on good terms with the scientific world. Then, just as the Empire was on the point of falling in 1814, Laplace deserted his Emperor, transferred his allegiance to the Bourbons, and was rewarded with a Marquisate.

Far-reaching though Newton's astronomical work had been, it had hardly scratched the surface of the vast problem pre-

* *A Short Account of the History of Mathematics*, p. 426.

sented by the motions of the heavenly bodies. Newton had treated the planets as perfect spheres, or even as points, and had usually assumed that they moved under the influence only of the sun's attraction, the moon only under that of the earth, and so on. It is true that he had sometimes contemplated other possibilities, but his principal results had been obtained on simplifying assumptions such as those just mentioned, which reduced each problem to its barest and crudest form. The problem of the heavens in all its intricacy called for a more refined and more subtle analysis, and at the age of 22, Laplace set himself the task of providing it.

In the next 17 years (1771–87) he solved many difficult problems of dynamics and astronomy, and then aspired to write a book which should 'offer a complete solution of the great mechanical problem presented by the solar system, and bring theory to coincide so closely with observation that empirical equations would no longer find a place in astronomical tables'. The final result was not one volume, but six. The *Exposition du Système du Monde*, which appeared in 1796, gave a broad general explanation of the results obtained. The larger *Mécanique Céleste* subsequently appeared in five volumes, four being published from 1799 to 1805 and a fifth, which was mainly historical, in 1825. These contain a fuller discussion of the problems of the solar system, though still without as much detail as most readers would like—for when Laplace had convinced himself of the truth of a result, he did not go to much trouble to help his readers to follow him.

It would be impossible here even to mention all the problems that Laplace treated; conspicuous among them were the tides of the oceans, the flattened shapes of the earth and other planets, and a variety of problems resulting from the gravitational pull of the planets on one another. Two special problems may be mentioned as being both intrinsically interesting and typical of the kind of subjects discussed.

The first formed the subject of Laplace's earliest scientific paper, which he contributed to the Académie des Sciences in

1773. The planets do not move round the sun in the perfect ellipses we should expect if the sun alone controlled their movements. The other planets act on them as well, and continually drag their orbits out of shape; we can imagine these orbits being so much altered that, for instance, the earth might ultimately become uninhabitable. Laplace attempted to show that there is no ground for such fears. This result was in direct opposition to one of the queries of Newton, who had thought* that the mutual action of planets and comets on one another would produce irregularities 'which will be apt to increase, till the system wants a Reformation' at the hand of its Creator.

Our second example comes from the *Système du Monde*. Laplace remarked that the planets all move in the same direction round the sun, and that their satellites all move in this same direction round their planets.† In the query we have just quoted, Newton commented on this regularity, and suggested that 'blind fate could never make all the planets move one and the same way in orbits concentric', and attributed the regularity to an order which had been introduced by the Creator and, as we have just seen, occasionally needed re-establishment at His hand. 'For it became him who created them to set them in order. And if he did so, it's unphilosophical to seek for any other origin of the world, or to pretend that it might arise out of chaos by the mere laws of nature....Such a wonderful uniformity in the planetary system must be allowed the effect of choice.'

Laplace agreed with Newton that the regularity was very unlikely to have resulted from mere chance, but held a different view as to its origin. Newton had postulated a Creator as the source of the regularity: Laplace 'n'avait pas besoin de cette hypothèse-là', as he said to Napoleon on

* *Opticks*, query 31, p. 402.

† So Laplace thought, but since his time all the four major planets—Jupiter, Saturn, Uranus and Neptune—have been found to possess satellites which do not conform to this rule.

handing him a copy of his book; he considered that the natural causes which had produced the planets could also produce the regularity. And he advanced his famous 'Nebular Hypothesis' as to the origin of the planets.

He imagined that the sun had begun as a nebulous mass of hot gas in a state of rotation. This gradually cooled, and as it cooled it shrank. The Newtonian mechanics required it to rotate ever faster and faster as it shrank. Laplace had shown that the orange-shaped flattening of the earth and planets resulted from their rotations—the faster a planet rotated, the flatter it would become. So he supposed that as the sun rotated ever faster, its shape became ever flatter until it assumed a disk-like shape. Then it could flatten no further, but broke into pieces by shedding ring after ring of matter from its protruding equator. He supposed that these rings of matter finally condensed and formed the present planets. These, too, would start as masses of hot rotating gas, and would go through the same series of changes as their parent sun before them; they too would cool, shrink, flatten in shape, and finally throw off rings of matter which would in time condense, thus forming the satellites of the planets. It was now easy to see why the planets and satellites were all revolving and rotating in the same direction; the direction was that in which the primeval sun had rotated.

This set of ideas was not entirely novel, for Immanuel Kant, the German philosopher, who had started life as a scientist, had propounded something very similar.* Laplace was probably unaware of this, for he said that he knew of no one except Buffon who had given any thought to the matter, and Buffon's thought had led him to the quite different theory that the planets had resulted from some astronomical object crashing into the sun and splashing out planets. In any case, Kant's hypothesis was inferior to that of Laplace in many ways; he had supposed, for instance, that the sun had acquired its rotation through shrinkage—a mathematical impossibility.

* *Allgemeine Naturgeschichte*, 1755.

At the time of its appearance, and for many years after, Laplace's theory was widely accepted as a plausible and interesting conjecture as to the origin of the planets and their satellites, but detailed discussion has now shown that it is untenable on at least two grounds.

The first is, in brief, that if the planets had been born through the ever-increasing rotation of a shrinking sun, then the present shrunken sun ought to be rotating very rapidly indeed, whereas in actual fact it is hardly rotating at all.*

The second criticism depends on properties of matter which were not known at the time of Laplace. If gas is let loose in space, it will tend to expand under its own internal pressure, and will usually end by scattering thoughout space. But if the amount of gas is substantial, the mutual gravitational attractions of its particles will tend to hold it together, and, for a large mass of gas, may even neutralise the universal tendency to expansion. Laplace's mass of hot gas would cool and shrink so slowly that it could only throw off matter in driblets. Calculation shows that the masses thrown off in this way would merely scatter through space, and so could not condense into planets. For these and a variety of other reasons, Laplace's Nebular Hypothesis has fallen completely out of favour.

Observational Astronomy

While all this was happening in dynamical astronomy, observational astronomy was having its own series of successes. The principal names we meet are Bradley (1693–1762), Sir William Herschel (1738–1822) and his son Sir John Herschel

* To state the criticism in a more precise form, the Newtonian mechanics requires that a certain mathematical quantity, the total angular momentum, should remain unaltered through all changes of the system. Theory tells us what value this quantity must have for the sun to break up through excess of rotation, while observation tells us the value it has now. The two are not equal, or even comparable, so that the sun cannot have broken up in the way that Laplace imagined.

(1792–1871), F. W. Bessel (1784–1846), F. G. W. Struve (1793–1864), U. J. J. Leverrier (1811–77) and J. C. Adams (1819–92).

BRADLEY (1693–1762). Bradley was Astronomer Royal at Greenwich when he discovered the important phenomenon of the 'aberration of light' in 1725. When we are in a moving boat on a perfectly windless day, the motion of the boat appears to create a head-wind. And if a wind blows up from the west it will not seem to arrive from a true westerly direction but from some other direction depending on the speed of the boat and the direction of its motion—there will be a sort of compromise between a head-wind and a westerly wind. If we change direction of the boat's motion, the wind will appear to change its direction at the same time. Bradley discovered an analogous effect in astronomy: as the earth changes the direction of its motion round the sun, the light from a distant star appears continually to change the direction from which it comes. Bradley found that all his observations could be explained by the single hypothesis that light travelled through space with a constant and finite speed, and this proved to be substantially the same as Römer had deduced from his observations on Jupiter's satellites (p. 199). On the whole it was Bradley rather than Römer who convinced the scientific world that light travelled with finite speed.

Bradley also discovered the phenomenon of 'nutation'. We have seen how Hipparchus had discovered the 'precession of the equinoxes'—the earth's axis of rotation does not point permanently in one direction in space, but moves round in a circle which it completes in 26,000 years. Bradley now found that the precise path was not a clean circle but a crinkly circle, the direction of the pole making a slight wobble every 19 years or so.

Later he measured the positions of a number of stars with great accuracy, and published a star catalogue in 1755 which proved to be of great value at a later date when the motions of the stars came under discussion.

HERSCHEL (1738–1822). Friedrich Wilhelm Herschel was born in 1738 in Hanover, which at that time belonged to the British Crown. He came to England as a professional musician, but after twenty-four years abandoned music for astronomy. His first interest was in the making of telescopes and the figuring of mirrors, in which he acquired great skill.

He was perhaps best known for his discovery of the planet Uranus in 1781. This made a great sensation at the time, since only five planets had been known from prehistoric times, and it had become ingrained in the human mind that the planets must be just five in number. Six years later he discovered that the newly discovered planet was accompanied by two satellites, and in 1789 he added two more satellites, Mimas and Enceladus, to the five which Saturn was known to possess.

Herschel's discovery of these new members of the solar system was important in its way, but his survey and study of the 'fixed stars' was of more fundamental and far-reaching importance. He made four complete surveys of the stars of the northern hemisphere with instruments of his own construction, and his son Sir John Herschel (1792–1871) subsequently extended these to the southern sky.

It was generally agreed by now that the sun is a star, similar to the other stars in structure, and a member of a great system of stars which is isolated in space and bounded by the stars of the Milky Way. Precision had been given to the idea by Thomas Wright (1711–86), an instrument maker of Durham.* Herschel's detailed study of the stars suggested that this system is shaped like a wheel or a millstone, and that the sun is near to its centre. The first of these conclusions has stood the test of time, but the second not; we now know that the sun is very far from the centre of the great wheel of stars.

Herschel could investigate the shape of this system of stars, but could not discover its dimensions, since the time of measuring the distances of the stars had not yet come. If the

* *An original Theory or new Hypothesis of the Universe* (1750).

VII

Herschel's 40-foot telescope

stars were all similar structures to the sun, all might be expected to be of the same intrinsic brightness as the sun; they would appear faint and bright simply because of their different distances from us, and a star's faintness would disclose its distance. For instance, on the supposition that Altair was of the same intrinsic brightness as the sun, Newton had estimated its distance to be about 5 light years, a light year being the distance that light travels in a year—about 5,880,000 million miles.* But the actual distance of a star was not measured until 1838, and it was still a long road from this to a measurement of the size of the galaxy. Recent investigations have shown that the diameter of the galaxy is of the order of 100,000 light years, the sun being at a distance of about 30,000 light years from the centre.

Herschel also made a prolonged series of observations on star clusters and nebulae, the latter being faint fuzzy-looking objects, of which only very few can be seen without powerful telescopic aid. Maupertuis described them as 'small luminous patches, only slightly more brilliant than the dark background of the sky; they have this in common, that their shapes are more or less open ellipses; and their light is more feeble than that of any other objects to be seen in the heavens'.† Maupertuis had conjectured that they might be immense suns, flattened by rotation. Kant the philosopher dissented from this view, saying:‡ 'It is much more natural and reasonable to assume that a nebula is not a unique and solitary sun, but a system of numerous suns which appear crowded, because of their distance, into a space so limited that their light, which would be imperceptible if each of them were isolated, suffices, owing to their enormous numbers, to give a pale and uniform lustre. Their analogy with our own system of stars: their form, which is precisely what it should be according to our

* We now know that it is nine times as intrinsically bright as the sun, and that its distance is 15 light years.

† *Discours sur les différentes figures des astres*, 1742.

‡ *Allgemeine Naturgeschichte und Theorie des Himmels*, 1755.

theory; the faintness of their light, which denotes an infinite distance; all are in admirable accord, and lead us to consider these elliptical spots as systems of the same order as our own —in a word to be Milky Ways.' This interpretation of these nebulous objects soon became a recognised astronomical speculation, and was known as the 'island universe' theory. It has stood the test of time, being in close agreement with our present knowledge (p. 353).

In 1781, Messier had published a list of 103 of the more conspicuous of the nebulae and similar objects. Herschel now set to work to extend this list, and his son carried it down to the southern sky. Their final list contained about 5080 objects, 4630 of which they had observed themselves. After being in general use for nearly a century, their catalogue was superseded by the still more comprehensive *New General Catalogue* of Dreyer (1890) and other lists of nebulae.

Herschel also studied the motions in the sky of the so-called 'fixed stars'. From the time of Aristotle and Ptolemy, it had been supposed that these did not change their positions except for their daily rotation round the pole. Then Halley had noticed in 1718 that Arcturus and Sirius no longer occupied the positions which Ptolemy had assigned to them, so that they must have moved relative to the other stars. Herschel now found that all the stars in one-half of the sky appeared to be scattering apart, while those in the other half all appeared to be closing in on one another. He concluded, rightly as we know, that the sun was moving towards the former region of the sky, and away from the latter. A detailed study of the motions (1805) showed that the sun is moving towards a point in the constellation of Hercules at a speed of about 13 miles per second. And if the sun was in motion in this way, it was natural to suppose that all the stars had comparable motions: they lost their right to be called 'fixed stars'.

Herschel also gave much thought to double and binary stars—pairs of stars which look close together in the sky. Some such pairs are bound to be, through one star lying nearly behind another when viewed from the earth, but the number of such cases is calculable from the laws of probability, and it had been found that it fell far below the number actually observed. Some of the pairs, then, must consist of stars which lie close together in space, probably because they are physically associated. Only time could show to which class any individual pair belonged; stars which were physically associated would be likely to stay near to one another, while stars which merely happened to lie on the same line of sight would be likely to move ever farther apart as they journed through space. In the years 1782–4 Herschel catalogued the positions of stars forming about 700 close pairs, and by 1803 he had found that many of them consisted of stars revolving round one another in just the kind of orbits that the Newtonian law of gravitation required. Such stars were presumably kept together by their mutual gravitational attractions, like the earth and moon, and so showed that gravitation was no merely local effect but extended throughout space. The study of such pairs of stars has led to valuable and interesting results. Until a few years ago, it was the only way of determining the masses of the stars; the astronomer calculates the gravitational pull needed to keep the stars of such a pair moving in their orbits, and from this, in favourable cases, he can calculate the mass of both stars. If the plane in which the stars move happens to pass quite close to the earth, the stars may be seen to eclipse each other as they sweep round in their orbits, and the duration of the eclipses may disclose the dimensions of the stars.

There is not space to record the steady growth of observational knowledge in the remainder of this period, but four landmarks are too important to pass by without notice.

The Asteroids. Early in the nineteenth century, on the evening of 1 January 1801, Piazzi discovered the asteroid 'Ceres', the largest of a family of minute planets which move round the sun in orbits between those of Mars and Jupiter. Well over a thousand of these objects are now known; they are conjectured to be the shattered fragments of what was originally a full-sized planet moving in the space in question.

The Distances of the Stars. The year 1838 provides our next landmark, for in December of that year the first reliable measurement of the distance of a star was announced by Friedrich Wilhelm Bessel at Königsberg. The star was 61 Cygni, and Bessel announced its distance to be 640,000 times the distance from the earth to the sun; modern measurements increase the figure to 680,000. At Dorpat, near St Petersburg, F. G. W. Struve had previously measured the distance of Vega (α Lyrae), but his measures, which had never inspired much confidence, ultimately proved to be more than double the true distance. In January 1839 Thomas Henderson, Her Majesty's astronomer at the Cape, announced measures of the distance of yet another star, α Centauri, but these have proved to be in error by about one-third.

In every case, the method was that by which a surveyor measures the distance of an inaccessible mountain peak or other landmark; he measures out a base-line, moves from one end of it to the other, and notices by how much the direction of the landmark changes as he does so. The astronomer takes the diameter of the earth's orbit for his base-line, the earth's motion taking him from one end of it to the other in six months. Knowing the angle by which the direction of a star differs at the two ends, it is easy to calculate the distance of the star in terms of the dimensions of the earth's orbit. Since 1838 immense numbers of stellar distances have been measured with ever-increasing exactness. New methods of greater range have also been developed, so that the dimensions of the universe can now be measured out with some accuracy.

The Discovery of Neptune. In 1846 another addition was made to the list of planets.

The planets do not describe perfect ellipses, irregularities being introduced by the attractions of the other planets, and the motion of Uranus had shown irregularities which could not be attributed to the attractions of the then known planets, and so were thought to be caused by some hitherto undiscovered planet. So at least it seemed to the young Englishman John Couch Adams (1819–92), who was subsequently Professor of Astronomy at Cambridge, and to the French mathematician and astronomer Urbain J. J. Leverrier (1811–77), and both men set to work to calculate the orbit in which an unknown planet would have to move if it were to produce the observed vagaries in the motion of Uranus. Adams finished his calculations first and asked the Cambridge observers to look for the planet at the position he had calculated for it. Leverrier subsequently did the same with the observers at Berlin. But as Berlin had better star-charts than Cambridge, the planet was first identified at Berlin. Yet another planet, Pluto, was discovered in a similar way by Tombaugh at Flagstaff Observatory, Arizona, in March 1930.

Astronomical spectroscopy. The last landmark, possibly the most important of all, is the application of spectroscopy to astronomy. Hitherto the stars had been nothing more than shining points of light; from now on they became giant crucibles in which nature performed experiments which we could watch; the spectroscope was to disclose their chemical composition, their temperatures and their physical states, as well as supplying a great variety of other knowledge.

The spectroscope is an instrument which analyses light into its constituent colours, which it then displays as a band of continuous coloured light known as a 'spectrum'; it is in fact an elaboration of the prism which Newton used for this same purpose (p. 205). The science of spectrum analysis was virtually founded in 1752, when Melvill noticed that a flame in which a salt or metal was being burned gave out a spectrum which consisted only of bright lines, the pattern of which depended on the substance which was being burned. In 1823 Sir John Herschel suggested that it might be possible to identify a chemical substance from an examination of its spectrum, and the road to the methods of modern spectroscopy lay open. In 1855 the American, David Alter, described the spectrum of hydrogen, and in the next few years the spectra of a great number of substances were identified by R. W. Bunsen (1811–99), G. R. Kirchhoff (1824–87), H. E. R. Roscoe (1833–1915) and many others. In 1861 Bunsen and Kirchhoff discovered the two new elements caesium and rubidium by spectroscopic methods.

The application of these methods to astronomy was slow in coming. In 1802, fifty years after Melvill's observation, W. H. Wollaston (1766–1828) had passed sunlight through a prism, as Newton had done before him, and obtained a coloured band of light which was not continuous but was interrupted here and there by a pattern of dark lines. The same lines were again noticed in 1814 by Joseph Fraunhofer, a Bavarian instrument-maker (1787–1826). Using the greater amount of light that was collected by a telescope, he was able to try the same experiment with the light from various stars, and obtained similar, but not identical, results. Again there were dark lines crossing the spectrum, but these were no longer in the same position as for sunlight, and, moreover, varied from star to star.

So far there was no obvious connection between terrestrial and stellar spectra; the former consisted of bright lines with dark spaces in between, while the latter consisted of dark lines with bright spaces of continuous colour in between. Connection was established in 1849, when Foucault noticed that a dark line in the spectrum of the sun, which Fraunhofer had designated as D, could be made brighter or darker by the use of the electric arc. If the sunlight which went into the spectroscope had been previously passed through an electric arc, the D line appeared darker than it did for ordinary sunlight, suggesting that there was something in the arc which could absorb the light of the D line. But there was also something which could emit this light, for if the sunlight was suddenly shut off, so that the light emitted by the arc was itself analysed without any sunlight, the same D line appeared, but now as *a bright line*.

Kirchhoff and Bunsen next found that the bright and the dark D lines both owed their existence to sodium. They put sodium in a flame, and found that the D line was brilliantly emitted. They passed the light from white-hot lime, which normally gave a continuous spectrum, through a flame of alcohol in which a small amount of sodium had been put, and found a dark line precisely in the position of the D line. They concluded that the substance which emitted the D line in the electric arc and absorbed it in the sun must be sodium. There must, then, be sodium in the sun.

Clearly the next step was to test the sun's spectrum for other chemical elements. They tried replacing the sodium by lithium, and obtained a different set of spectral lines which they could not identify with any of the Fraunhofer lines, so they concluded that there was little or no lithium in the sun. Here was the beginning of a technique for the study of the chemical composition of the stars.

A vast number of investigators now embarked on the study of stellar spectra, Sir William Huggins (1824–1910), Janssen (1824–1907) and Sir Norman Lockyer (1836–1920) being conspicuous among the pioneers. Most of the lines which they found in the spectra could be ascribed to substances which were already known on earth, and it was assumed that these same substances entered into the composition of the stars. But this was not true of all the lines of the spectrum. In 1868 Lockyer found a line in the solar spectrum which he could not attribute to any known source, and supposed that it must come from some new element, hitherto unknown on earth. The new element was called helium, and was found in 1895 to be a normal constituent of the earth's atmosphere.

Such were the beginnings of the now important science of spectroscopy, to which we shall return in the next chapter.

OPTICS

After the theories of Newton and Huygens had equally failed to account for the double refraction of Iceland spar (p. 211), the subject of optics lay quiescent throughout the eighteenth century, but sprang to life again with the coming of the nineteenth. For in 1801 Young discovered the hitherto unnoticed property of wave-motion that he described as 'interference', and this was found capable of solving most of the outstanding puzzles of optical theory in terms of a purely undulatory conception of light.

THOMAS YOUNG (1773–1829) was a conspicuous instance of an infant prodigy who retained great and varied abilities in mature life, in which he achieved distinction as physicist, physician, mathematician, linguist, philologist, antiquarian and scholar. It is said that he was able to read with fluency at the age of two, and could repeat Goldsmith's *Deserted Village* by heart before he was six.* By the age of 19 he had acquired a thorough knowledge of Latin and Greek, and a considerable acquaintance with Hebrew, Chaldee, Arabic, Syriac, Persian, French, Italian and Spanish—an array of linguistic accomplishments which proved their worth later in life when he became interested in Egyptology, and finally made a most valuable contribution towards deciphering the Rosetta stone in 1815. He decided to study for the medical profession, qualifying as a Doctor of Physic in 1796. He practised as a physician for a large part of his life afterwards, but first he founded the science of physiological optics. We have seen how Kepler had thought that long and short sight resulted from a want of accommodation in the crystalline lens of the eye. In 1793 Young proved conclusively that accommodation of the eye for vision at different distances resulted from changes in the curvature of this lens. He read a paper on this subject to the Royal Society in May 1793, and was

* *D.N.B.*, art. Thomas Young.

elected a Fellow of the Society a year later—immediately after his 21st birthday.

In 1798 he performed some experiments on light and sound, which he described to the Royal Society in November 1801.* To explain his results, he imagined a sequence of ripples moving across a lake and continuing along a narrow canal leading out of the lake. Suppose that after some distance this

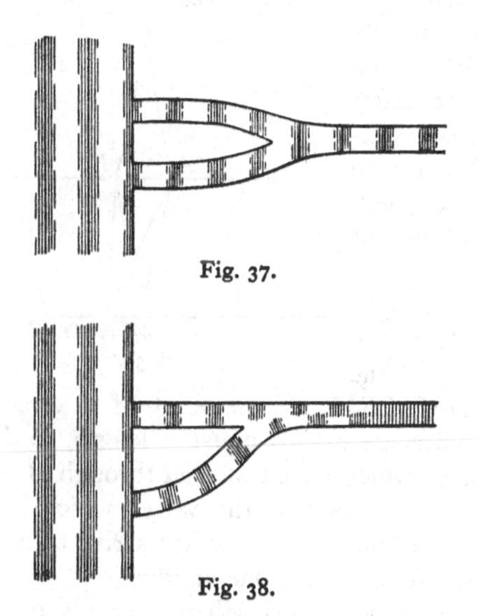

Fig. 37.

Fig. 38.

canal is joined by another which comes out of the same lake and has similar waves travelling along it. On the part beyond the junction, the two sets of waves will, of course, become mixed up, and will travel on together.

If the two canals are precisely similar, the two sets of waves will be precisely similar when they meet, and so will simply pile up into waves of double height (fig. 37). But if the canals are of different lengths, the crests of one set of waves may happen just to coincide with the troughs of the other set when

* 'On the theory of light and colour', *Phil. Trans.* 1801.

they meet, and just fill these up; in this case the surface of the water will remain entirely smooth. Young describes the mutual destruction of the two sets of waves as 'interference', saying 'I maintain that similar effects will take place wherever two portions of light are thus mixed, and this I call the general law of the interference of light.'

The theory is both illustrated and tested in one of Young's experiments. He caused light from a source *L* (fig. 39) to fall on a card in which two minute pinholes, *A*, *B*, had been punched close together. After passing through these pinholes, the light fell on to another screen *MN* beyond. The two pinholes of course represent the two channels of Young's explanation, and light can reach any point *M* on the lower screen by either of the alternative routes *LAM*, *LBM*. If the point *M* is directly under *L*, these two paths are exactly equal in length, so that the crests of the waves which have travelled through *A* will coincide exactly with the crests of the waves which have travelled through *B*. The illuminations which arrive through *A* and *B* will accordingly reinforce one another.

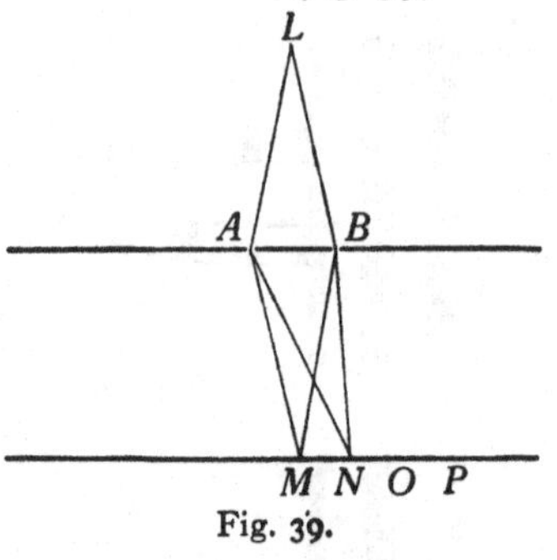

Fig. 39.

At another point such as *N*, this will not usually be so, since the paths *LAN* and *LBN* are not generally equal. If their lengths happen to differ by just a complete wave (a 'wave-length') of the light, then crest will again fall on crest, and again the waves will reinforce one another. The same thing will happen if the difference of length is any exact number of wave-lengths. But if the difference of path is just half a wave-length, then crest will fall on trough, and the two sets of waves will neutralise one another by interference. There will now be no light at *N*, but complete darkness. The same will happen if the difference of path is one and a half wave-lengths or is equal to any exact number of wave-lengths plus half a wave-length. Thus we may have full illumination at *M*, complete darkness at *N*, full illumination again at *O* and so on, giving a regular alternation of light and darkness which is described as an interference pattern. Young found that all these predictions of theory were confirmed by experiment, and many others as well.

All this seems simple and convincing enough to-day, but its importance was not understood at the time, and Young's work was first condemned and then ignored, one critic of position blaming the Royal Society for printing such 'paltry and unsubstantial papers' which he considered to be 'destitute of every species of merit'.

About 14 years later, Fresnel took up the subject, developing it with great mathematical skill, and showing that the theory of wave-motion could account for all the then known phenomena of optics, including the double refraction of Iceland spar, which had formed such a stumbling-block to the earlier theories of Newton and Huygens. From now on, it was generally agreed that light must consist of undulations, since it seemed impossible that two streams of particles could neutralise one another in the way observed. It was usual to think of the undulations as occurring 'in the ether', but the question of what the ether was, and how it undulated, were to remain unanswered for a long time (p. 287).

The experiments we have described were performed with monochromatic light, i.e. with light of one single spectral colour. If white light is used, the different constituent colours of the light are of different wave-lengths, and so produce different-sized patterns on the screen, the result being a complicated pattern of vivid colours. From the dimensions of the pattern formed by any single colour of light, it is possible to deduce the wave-length of the light. Fresnel found in 1821 that red light has about 40,000 waves to the inch, and violet light about 80,000, the intermediate colours being of course intermediate in wave-length.

Remote though the two subjects may seem, this success of the undulatory theory was intimately connected with Fermat's principle that the time of travel of a ray of light attains its minimum value along the path actually travelled by the ray. At any point on this path we may, as Huygens proposed, imagine that innumerable wavelets are arriving by different paths. As the time of travel along the actual path is a minimum, the time of travel along any immediately adjacent path must be substantially the same,* so that when the various sets of waves mingle, crest will just fit into crest, and the sets of waves will reinforce one another. Thus there will be an accumulation of light along the path of minimum time, but not elsewhere, and Fermat's principle is seen to be a direct and inevitable consequence of the undulatory theory of light.

In 1849, H. L. Fizeau measured the velocity of light in air and found a speed of about 315,300 km. per second. The next year his compatriot, J. L. Foucault, started a prolonged series of experiments by a different method, and obtained the much more accurate value of 298,600 km. per second. By modern experiments of great refinement, Michelson has measured the velocity as about 299,770 km. (186,300 miles) per second. Foucault also proved that light travelled less rapidly in a denser medium, the diminution of speed being just that required by Fermat's theorem and the undulatory theory of light. This drove a final nail into the coffin of the corpuscular theory of light, which could only explain the refraction of light by assuming that light travelled *more rapidly* in denser media.

* Because of the general principle that as a continuously varying quantity passes through its minimum value, its rate of change becomes momentarily zero—the passage from decrease to increase must be through a standstill at which there is neither decrease nor increase. Kepler had been acquainted with this general principle, but had never stated it clearly. It was given complete precision by the differential calculus which states that when a variable attains its maximum or minimum value, its differential coefficient has a zero value.

VIII

Henry Cavendish

THE STRUCTURE OF MATTER

In his *Opticks* (1704) Newton published his mature reflections on a number of questions which had no very direct connection with optics, and in particular on the structure of matter, to which he seems to have given a large amount of thought. In the last query (Qu. 31) of the optics, he writes: 'All these things being considered, it seems probable to me that God in the Beginning formed Matter in solid, massy, hard, impenetrable, moveable Particles, of such Sizes and Figures, and with such other Properties, and in such Proportion to Space as most conduced to the End for which he formed them; and that these primitive Particles being Solids, are incomparably harder than any porous Bodies composed of them; even so very hard, as never to wear or break in Pieces; no ordinary Power being able to divide what God Himself made one in the first Creation.' This was a direct return to the atomism of Democritus and Gassendi, and was probably fairly typical of the views held at the beginning of the eighteenth century.

In the same query, Newton wrote: 'Now the smallest Particles of Matter may cohere by the strongest Attractions, and compose bigger Particles of weaker Virtue, and many of these may cohere and compose bigger Particles whose Virtue is weaker still, and so on for divers Successions....If the Body is compact, and bends or yields inward to Pression without sliding of its Parts, it is hard and elastic, returning to its Figure with a Force rising from the mutual Attraction of its Parts. If the Parts slide upon one another, the Body is malleable or soft....Since Metals dissolved in Acids attract but a small quantity of the Acid, their attractive Force can reach but to a small distance from them....'

This last sentence introduces us to another view which was commonly held at this period—that the properties of matter, and especially its chemical properties, could be explained in terms of short-range forces which acted between the ultimate particles of the matter. This had been suggested by Boyle in 1664, by Hooke in 1665, and by Newton in the Preface to his *Principia* (1687) in words we have already quoted (p. 191). Huygens had said much the same thing in 1690: 'In true philosophy, the causes of all natural phenomena are conceived in mechanical terms. We must do this in my opinion, or else give up all hope of ever understanding anything in physics.'*

Here, then, was a school of thought which pictured the whole universe as mechanical, and all natural phenomena as evidence of forces acting on matter. But another school of thought imagined that matter did not consist solely of 'solid, massy particles', but contained also a variety of 'imponderables', perhaps also in the form of minute particles, and that these were responsible for many of the qualities with which matter could be affected.

For instance, a body was supposed to be combustible because it contained 'phlogiston', a term which had been introduced into science in 1702 by George Ernest Stahl (1660–1734), who subsequently became physician to the King of Prussia. The more phlogiston a body contained, the more readily it burned, but as it burned the phlogiston went out of it; this explained why a body changed its quality on being burned, and why the same body could not be burned twice. Phlogiston was in fact a sort of descendant of the 'sulphur' which the Arabian alchemists had taken to be the principle of fire. Indeed, sulphur was now thought to consist of almost pure phlogiston, because it left almost no residue when it was burned. Yet if this had been so, the weight of a body would have decreased on combustion, whereas Rey (p. 214) and Boyle (p. 213) had both shown that it increased.

Another supposed imponderable was 'caloric', which permeated the pores of bodies, and made the bodies hot or cold according to whether much or little of it was present. This was pressed into service to explain many of the phenomena

* *Traité de la Lumière*, chap. 1.

of heat—for instance, a body which was hammered became hot because the hammering squeezed the caloric to the surface. There were also the corpuscles by which Newton had tried to explain light (p. 209), and the 'electric fluids' or 'humours' or 'subtile effluvia' by which Gilbert had tried to explain electrical attractions and repulsions. Finally, there was the ether or other medium which filled the whole of space, transmitting light as well as serving innumerable other purposes. Kepler had used it to explain how the sun kept the planets in motion, Descartes to provide the basic material for his vortices, and Gilbert to account for magnetic phenomena; in general, it provided a convenient explanation for phenomena which were not easy to explain otherwise.

Gradually hard facts of experiment drove all the imponderables out of existence except the ether; this, in its special capacity of a transmitter of light-waves, lingered on into the early years of the present century. The story of the elimination of the other imponderables forms a large part of the history of eighteenth-century science.

Pre-eminently, however, the eighteenth century was the period in which the commonest substances, such as air and water, were studied and their chemical composition determined, while the commoner elements and the simpler gases were isolated and their properties recorded. We of to-day take it for granted that water is a 'chemical compound' of the two 'elements', 'hydrogen' and 'oxygen', and that air is a 'mechanical mixture' of the 'molecules' of the two 'gases', 'nitrogen' and 'oxygen'; not only were our scientific predecessors of 200 years ago ignorant of this, but none of the words printed in quotation-marks had any precise meaning for them. The word 'gas', which now first came into common use, a corruption of the Greek word 'chaos', had been introduced as far back as 1644 by the Flemish chemist, Jan Baptist van Helmont (1577–1644). He had recognised that there must be many kinds of gas, but he could not study their properties since he had no means of isolating them.

BLACK. The first step along the road was taken by the Scottish chemist Joseph Black of Edinburgh (1728–99). In 1756* he proved the existence of a gas, different from air, which could be made to combine with a variety of substances, but could also exist in the uncombined state. It was carbon dioxide, but he called it 'fixed air', because it could be fixed by combination with other substances.

CAVENDISH. The next advance was made ten years later by the strange genius Henry Cavendish (1731–1810), who, notwithstanding his many eccentricities, was one of the greatest scientists of his age. He was a tragic mass of inhibitions and complexes, so shy that he could hardly bring himself to speak to anyone, especially of the other sex, and then was able to speak of nothing outside his science. Thus he cut a ridiculous and pathetic figure in life, and even in death. After two days' illness in bed, he announced to his servant that he was about to die, and instructed the man to keep out of his way for some hours. The first time the servant re-entered the room, Cavendish snarled to him to keep out; the second time he found his master dead.

In 1766 Cavendish announced the discovery of another new gas which he described as 'inflammable air'. He produced it by the action of acids on metals, and investigated many of its properties. It was, of course, hydrogen. As with the 'fixed air' of Black, the choice of a name shows how all gases were at this time regarded as mere modifications of atmospheric air.

PRIESTLEY. Within the next few years a number of other gases were discovered by the Yorkshireman Joseph Priestley (1733–1804), who, in most respects except scientific ability, was the exact opposite of Cavendish. Cavendish had been born in the purple, with an ancestry of dukes; Priestley was the son of a cloth-finisher and a farmer's daughter. Cavendish died worth a million pounds which he had not known how to

* *Experiments upon Magnesia Alba, Quicklime and some other alcaline Substances.*

spend; Priestley was poor, often grindingly poor, and yet managed to make for himself a fairly happy and contented life, in marked contrast with the infinitely lonely and unhappy life of Cavendish. Cavendish had no interest in life outside his science, but Priestley was a man of many interests and hobbies, science being one, for by profession he was a minister of religion—first a Presbyterian and then a Unitarian.

His first discovery in chemistry was 'that exceedingly pleasant sparkling water' which we now call soda-water, and which was of course simply Black's 'fixed air' put into water under pressure. Priestley thought it might be a cure for scurvy, and persuaded the Admiralty to try it out, which they did by fitting two warships with soda-water generators. Among his more serious achievements, the rediscovery of oxygen in 1774 (p. 262) must certainly stand first.* He came upon it while examining the effects of heat on various substances which he floated on mercury inside a closed vessel, and then heated with the rays from a burning glass. He found that when oxide of mercury was treated in this way, it gave off a new gas in which flames burned more brightly and respiration was more vigorous than in ordinary air. As far back as 1771 he had noticed that air which was no longer able to support combustion or respiration regained its powers if green plants were inserted into it, even a sprig of mint or a piece of groundsel being effective. It was now easy to understand why—the new gas must obviously be essential to combustion and respiration, and was exuded by green plants. Its properties identified the new gas with that which Hooke and Boyle had postulated in 1662 and Mayow in 1674, and which Borch had prepared from saltpetre in 1678.

* Wilhelm Scheele, a Swedish apothecary, had published a *Treatise on Air and Fire* in 1777, in which he proved that air consists of a mixture of the two gases that we now call oxygen and nitrogen. As he had performed most of his experiments before 1773, he may have come upon oxygen before Priestley did so in 1774, but there is no question that Priestley was the first to publish the discovery.

Priestley called it 'dephlogisticated air', because it seemed to be greedy for phlogiston and to grab it when the two came into contact.

Priestley also discovered nitric and nitrous oxides, hydrochloric acid gas, sulphur dioxide and several other gases. He had the habit of collecting his gases through mercury instead of through water as others did, and in this way came upon gases which others overlooked through their being soluble in water.

Priestley was also responsible, although very indirectly, for the discovery that water is a composite substance, and not a simple chemical element. On one occasion, 'to please a few philosophical friends', he had exploded a mixture of common air and hydrogen by passing an electric spark through it, and noticed that the walls of the containing vessel became covered with moisture. Cavendish subsequently repeated the experiment, examined the nature of the moisture, and found it was pure water. Following this clue, he found that water consisted of Priestley's 'dephlogisticated air' and his own 'inflammable air' in the ratio of about 195 to 370 by volume, or, near enough, in the ratio of one to two. He clinched his proof in 1781 by producing water synthetically from a mixture of the two gases.

Priestley's discovery of oxygen did not, as might have been expected, give a death-blow to the phlogiston theory, and science continued to be encumbered with this and other imponderables, as well as with the relics of still earlier ages, including the four 'elements' of Empedocles, and the four primary qualities from which they were supposed to be compounded. For instance, the Empedocleans had said that earth was a mixture of cold and dry, while water was a mixture of cold and wet. If, then, wet could be changed into dry, water ought to be changed into earth. And, even in the eighteenth century it was still believed that water could be changed into earth by the simple expedient of boiling it. Newton illustrated his idea that 'Nature seems delighted

A metabolism experiment by the Lavoisiers

Drawing by Madame Lavoisier showing her husband at work, and herself taking notes, in the Arsenal Library

with Transmutations' by telling us that 'Water by frequent Distillations changes into fixed Earth'.*

LAVOISIER. This and similar relics of bygone ages were finally dispelled by the experiments of the French chemist Antoine Laurent Lavoisier (1743–94), who has a better right than most claimants to the title of father of modern chemistry, since in his hands chemistry really adopted modern methods and ideas. Born in Paris into a legal family of standing, he was sent to study law, although he had done particularly well at school in science. However, at the early age of 21 he gained a prize from the Académie des Sciences for an essay on street lighting, was elected to the Académie two years later, and became a full ordinary member in the succeeding year. But, like his great contemporary Laplace, he could not be content with a purely scientific life.

Within a few weeks of his election to the Académie, he became a member of the 'fermiers généraux', the intensely unpopular body of men to whom the government farmed out the collection of taxes, and from now on he divided his life between the two worlds of science and of business of a somewhat unsavoury type. At the age of 28 he married the daughter of another fermier, who brought him youth and beauty, wealth and brains, as well as sympathy and active help in his scientific career; it seems to have been a completely happy marriage.

When the Revolution came, there was intense feeling against the fermiers généraux, who had one and all made themselves rich at the public expense; they were all arrested, tried, and condemned to the guillotine. Attempts were made to get Lavoisier exempted in recognition of his scientific eminence and his services to science, but the general unpopularity of the fermiers pressed too heavily on the other side of the balance and the judge, remarking that 'the Republic had no need of savants', decided that the law must take its course. So died the greatest chemist of his time, at the age of 51.

* *Opticks*, query 31.

Since the time of Van Helmont (p. 257), many chemists had shown that distilled water, when evaporated in a glass vessel, always leaves a slight earthy deposit. The deduction was that water had been changed into earth. In 1770 Lavoisier showed this to be untrue. He heated water for a long time in a sealed glass vessel. A white solid slowly separated in the water but there was no change in the total weight. On pouring out the water and evaporating to dryness it was found that the total weight of solid matter formed was almost exactly equal to a loss in weight of the glass vessel which had occurred during the experiment. The conclusion was that the 'earth' had been dissolved by the water from the glass of the vessel.

He next (1774–78) attacked the prevalent idea that mass could be changed by chemical action. When metals were calcined they increased their mass, but Lavoisier showed that this was only because they abstracted something from the air; when this was adequately allowed for, the total mass showed no change. He showed that the same was true of respiration, which he treated as a slow type of combustion. In this way he introduced the idea of mass as something permanent and indestructible, something which was conserved through all changes. Newton had introduced the idea of a mass which remained constant through all *dynamical* change (i.e. through all change of motion); Lavoisier now showed that the same was true of chemical change.

Lavoisier had worked with oxygen before its discovery had been announced by Priestley, but it seems likely that Priestley told him how to obtain it from red lead when he met the Lavoisiers in Paris in October 1774.* Priestley's first public announcement of his discovery was in a letter to the Royal Society which was dated 15 March 1775, and was read a week later. Lavoisier was charged with stealing, but in 1782 he referred to oxygen as 'the air which Mr Priestley discovered about the same time as myself and even, I believe,

* *Three Philosophers* (*Lavoisier*, *Priestley and Cavendish*), W. R. Aykroyd, 1935.

before me'. Lavoisier was the first to realise the importance of oxygen and that it was an element. By an ingenious series of experiments he established the true theory of combustion and showed that the phlogiston theory was untenable.

He now embarked on a study of the chemical composition of various substances. He began with the simplest—air and water—and was able to show in 1777 that 'the air of the atmosphere is not a simple substance, an Element, as the Ancients believed, and as has been supposed until our own time', and he gave the proportion of its two constituents as three to one, the true proportion of nitrogen to oxygen being about 78 to 21. Seven years later he repeated the experiments of Priestley and Cavendish on the composition of water, confirmed their conclusions, and gave their present names to the two constituent gases—hydrogen (water-producing) and oxygen (acid-producing).

He then went on to analyse more complex substances into their elements, adopting Boyle's definition of an element as a substance which is so simple that it cannot be further decomposed. He divided the elements so obtained into four classes. The first consisted of the gases hydrogen, oxygen and nitrogen and the imponderables light and caloric; the second consisted of elements which formed acids when oxidised; the third consisted of metals; the fourth consisted of various earths, such as lime and magnesia, which Lavoisier mistook for elements because he could not further decompose them.

RUMFORD. Phlogiston had disappeared from the scheme of things when Lavoisier had shown that mass was conserved without phlogiston being taken into account. Caloric lingered until nearly 1850, although partially dismissed through the efforts of an American, Benjamin Thompson (1753–1814), who was better known as Count Rumford of Bavaria, a title he had acquired after administering the Kingdom of Prince Maximilian of Bavaria for many years. He was something of a wanderer and even adventurer, but finally settled down and married the widow of Lavoisier, though only for a brief time, since the marriage proved as unhappy for her as her previous marriage had been happy.

Rumford's scientific interests were mostly of a highly practical kind. He is generally credited with the invention of steam-heating and of the modern chimney, which apparently smokes less than its pre-Rumford predecessors used to do. In 1798 he published an *Inquiry concerning the Source of the Heat which is Generated by Friction*, in which he remarked that in boring cannon, the metal becomes very hot, there being virtually no limit to the amount of heat that can be produced. Ordinary heating, such as that of a kettle over a fire, had usually been explained as a transfer of caloric, but no such explanation was possible here, since there seemed to be no limit to the amount of heat which would have to be transferred. Rumford argued that it was more likely that heat was 'a kind of motion'.

He tried to test this conjecture experimentally. A good deal of heat is needed to melt ice into water, so that if the caloric theory were true, ice could only change into water by the acquisition of a large amount of caloric, with a corresponding gain of mass. But Rumford found in 1799 that the water produced by melting ice weighed just as much as the original ice, at least to within one part in a million. Again he concluded that heat must be weightless, and so could not be a substance but must be 'an intestine or vibratory motion of the constituent parts of heated bodies'. This experiment, however, did not disprove the theory that caloric was a weightless fluid and the caloric theory finally fell only owing to the success of the concept of heat as a form of energy.

Nineteenth-Century Chemistry

By the beginning of the nineteenth century, most serious scientists had dropped phlogiston out of their vocabularies, and had come to recognise that combustion was simply a process of combining with oxygen, and heat nothing more than a motion of the ultimate particles of matter. The commoner substances had been isolated and studied, while Black, Cavendish, Priestley, Lavoisier and others had shown that quantitative chemical measurements could be made with good accuracy. The road was now open to an exact study of the laws of chemical combination—why and how did substances combine to make other and new chemical substances?

PROUST. The first great stride along this road was made by the French chemist Joseph Louis Proust (1754–1826). He carried out a long series of experiments on the composition of a number of minerals and also on many metallic compounds produced artificially. He showed that the composition was constant and also that some metals formed more than one oxide and sulphide, each with a definite composition. This led to his formulating his 'law of definite proportions': 'In all chemical compounds the different constituents always enter in unvarying proportions.' This was verified for many other compounds by J. B. Richter (1762–1807) and E. G. Fischer (1754–1831) and others, and tables of the proportions drawn up.

DALTON. Another great advance was now made by the Quaker John Dalton (1766–1844), the son of a Cumberland weaver, who taught in a Manchester school and studied science in his spare time. Dalton wondered why the heavier and lighter gases in the atmosphere did not separate out, as oil and water do, and concluded that the constituent gases must exist in the form of small particles or atoms of the kind which had been imagined by Leucippus and Democritus (p. 44), and that these must be completely mixed together in the atmosphere.

This threw a new light on the laws of definite proportions. It was only necessary to suppose that the minute atoms could combine into small groups of uniform structure and so form more complex substances, and the mystery of this law was completely solved. Dalton suggested, for instance, that carbon monoxide is formed by atoms of carbon and oxygen pairing one with one, while carbon dioxide resulted from a single atom of carbon uniting with two atoms of oxygen.

If this were so, then the definite proportions of Proust's law would, of course, reveal the relative weights of the different kinds of atoms. Carbon dioxide, for instance, was known to consist of 3 grams of carbon to 16 grams of oxygen; if, then, it contained twice as many atoms of oxygen as of carbon, the weights of the atoms of carbon and oxygen must stand in the ratio of 3 to 8.*

PROUT'S HYPOTHESIS. When the relative weights of many atoms had been determined in this way, the striking fact emerged that the weights of most atoms were exact multiples of the weight of the hydrogen atom, or very nearly so. The English physician William Prout (1785–1850) drew attention to this in 1815, and suggested that all matter might consist simply of hydrogen atoms. For instance, the oxygen atom, with 16 times the weight of the hydrogen atom,† might consist of 16 atoms of hydrogen which had somehow become united together. To the question of Thales 'What is the fundamental substance of the universe?' Prout suggested the answer: hydrogen.

If the constituents of a compound are all gaseous, the proportion in which they combine may be measured by volume; it is found, for instance, that one volume of oxygen combines with exactly two volumes of hydrogen to form water, and in 1808 the French chemist Joseph Louis Gay-Lussac (1778–1850) announced that there is always a simple numerical relation of this type. The relations, however, can be put in still simpler forms; from that just mentioned we can deduce that under similar conditions a number of atoms of oxygen occupy the same volume as an equal number of atoms of hydrogen.

This is only one case of the simpler and far more general

* Dalton gave this and a number of other instances of his theory, the given numbers often being inaccurate, but the principles completely sound.

† Actually Dalton had said 7 times, his error arising in part from his thinking that the water molecule consisted of a single atom of hydrogen combined with a single atom of oxygen.

law which is generally known as Avogadro's Law: 'Conditions of pressure and temperature being the same, a given volume of gas will always contain the same number of molecules, whatever the nature of the gas may be.' In 1811 the Italian chemist Amedeo Avogadro had seen that some simple law of this general type must be implied in the laws of Gay-Lussac and of Dalton, but for a long time there was some doubt as to the exact relation, until finally the Italian chemist Stanislao Cannizzaro (1826–1910) cleared the matter up in 1858. Avogadro had already introduced the term 'molecule' to denote the small groups into which atoms combine, and Cannizzaro showed that the true form of the relation was that just stated. It now became important to discover the number of molecules in a gas at any assigned temperature and pressure. In a gas at standard atmospheric pressure and temperature the number per cubic centimetre is found to be about $2{\cdot}685 \times 10^{19}$, a number which is usually known as Loschmidt's number.

It had long been clear that the physical properties of the elements were not mere random assortments. On the contrary, the elements fall into groups in which all the members possess similar, although not identical, properties. There are, for instance, groups of the metals, and again the group of monatomic gases (discovered 1894–1908) in which the atoms are so inert that they do not form compounds with atoms of other substances, or even with one another. Chemists now began to wonder how the physical properties of the various elements were related to their atomic weights. The problem was especially studied by Lothar Meyer (1830–95) and Dmitri Mendeleef (1834–1907), the latter of whom produced the 'periodic table' which figures so prominently in modern chemistry.

It might perhaps be anticipated that the elements of high atomic weight would exhibit one group of properties, those of low atomic weight another, and so on; but the periodic table shows something very different. If the elements are arranged in order of increasing atomic weight, and numbered

1, 2, 3, etc., then the elements numbered 2, 10, 18, 36, 54, are found to show similar properties; again those numbered 3, 11, 19, etc., show similar properties, although different from those shown by the former group, and so on; the properties repeat themselves periodically. There is a physical reason for all this, as we shall see later, but it took a long time to find it out.

Energy and Thermodynamics

The problem of the structure of matter had now been put on a sure foundation, so that it was possible to turn to problems dealing with the behaviour of matter.

CARNOT (1796–1832). In 1824 a French engineer Sadi Carnot issued the one and only publication of his scientific life—*Réflexions sur la Puissance motrice du Feu.* It was a remarkable paper on the theoretical side, for it not only founded the modern science of thermodynamics, but also gave it its present form. But Carnot cared very little about the theoretical side of his problem; for him it was a problem of industrial economy, which he approached as a practical engineer. Steam engines, which were now coming into common use, had to be fed with fuel, and did work in return. Carnot wanted to know how much work could be obtained for a given expenditure of fuel, and above all he wanted to know how to make this amount of work a maximum. He thought of work in terms of the raising of a weight through a height, and his discussion turned on the conception of heat being interchangeable with work.

JOULE (1818–89). Carnot's ideas had no great influence on the progress of science until they were recaptured by James Prescott Joule of Manchester, a pupil of Dalton. Joule performed a series of very skilful experiments on the relation of work and heat, measuring the amount of heat generated by a known amount of work expended, for instance, in stirring water round in a vessel. He defined his unit of work, the 'foot-pound', as the work done in raising a pound weight through a foot, and his unit of heat as the amount of heat which raises a pound of water through a degree on the Fahrenheit scale. He then showed* that there is a fixed relation between the two units, so that one unit of heat always yields up the same definite number of units of work—the number which we now describe as the 'mechanical equivalent of heat'. It must already have been a commonplace that heat and work were somehow interchangeable; Joule showed experimentally that they are interchangeable *at a fixed rate*, and determined what the rate is. Work could also be performed by other agencies than heat, as, for instance, by electrical and radiant energy. There are also kinetic energy of motion, which was usually measured as *vis viva* (p. 234), but which Joule described as 'living force', and potential energy, such as the energy of a raised clock-weight. Joule found that these, too, were interchangeable at fixed rates, each having its exact equivalent in terms of heat or work. In 1847 he announced that 'Experiment has shown that when living force is *apparently* destroyed, whether by percussion, friction, or any similar means, an exact equivalent of heat is restored. The converse is also true....Heat, living force and attraction through space (to which I might also add light, were it consistent with the scope of the present lecture) are mutually convertible. In these conversions, nothing is ever lost.'†

Not only was nothing ever lost, but nothing was ever gained, since all conversions took place at a fixed rate of exchange. This showed, in one vivid flash, why perpetual motion was an impossibility; every natural system contained a definite but limited capacity for doing work, while perpetual motion called for an indefinite and unlimited capacity.

The argument could be turned the other way round. If experience shows that perpetual motion is impossible, this

* *On the Calorific Effects of Magneto-electricity and on the Mechanical Value of Heat*, 1843.

† This was first announced to the world in a popular lecture in a Church reading-room, and was first published to the world in a Manchester weekly paper.

must mean that no power of doing work can ever be gained by transformations, and hence that all transformations must be at fixed rates of exchange. In 1847, the same year in which Joule announced his general conclusion, the German physicist Hermann von Helmholtz (1821–94) published a small pamphlet, *Erhaltung der Kraft*, in which he adduced consideration of this kind, and reached the same conclusion as Joule: heat and work were interchangeable *at a fixed rate.* Joule had found that this was so by direct experiment, while Helmholtz had seen that it must be so by an abstract argument based on the impossibility of perpetual motion.

LORD KELVIN (1824–1907). About this same time Sir William Thomson, afterwards Lord Kelvin, began to study these doctrines mathematically, and to build on them a consistent body of knowledge. It had now been shown that every material system contained a pool of heat, 'living force', etc., which represented capacities for doing work, and could change into one another at fixed rates of exchange, the total remaining the same through all transformations, except in so far as it was increased or diminished from outside. Kelvin described this as 'energy', a term which had been introduced into physics by Young.

Joule's principle could now be expressed in the form that energy was conserved. His first enunciation of the principle had been concerned primarily with mechanical forces and heat, but he had shown that the principle applied equally to electrical energy. In 1853 the chemist Julius Thomsen found that energy is conserved in chemical transformations also. In this way the wide general principle of the conservation of energy became established in physical science, and formed a fitting companion to the conservation of mass. Energy, like mass, was seen to be of constant total amount, all the changes of the universe arising from redistributions of pools of energy and mass which never change their total amounts.

Lord Kelvin now proposed a precise scale for the measurement of energy or heat. Accepting that heat was a random motion of the particles of a body, he proposed that temperatures should be measured from a zero-point at which there was no such motion—the 'absolute zero' of temperature. Certain theoretical considerations advanced by Carnot showed that this would be the same for all substances; experiment fixed it at $-273°$ C. Thermometers had hitherto depended on the expansion of mercury, or of some other special substance, but Kelvin's new 'absolute scale' was independent of the properties of special substances.

THE KINETIC THEORY OF GASES

It was now possible to inquire how the molecules and atoms of a substance must function, so as to endow the substance with its observed physical properties.

The problem was simplest for a gas. Boyle had tried to explain his 'spring of air' (p. 215) by comparing air to 'a Heap of little Bodies, lying upon one another, as may be resembled to a fleece of wool. For this consists of many and flexible little Hairs which, like Springs, can be curled up, but will always try to uncurl themselves again'.* This implied that the particles of air must be in contact with one another. Gassendi,† on the other hand, had thought that they must be spaced far apart, and kept far apart by their motion. Such a picture, he thought, would account for all the physical properties of a gas.

Twenty years later Hooke advanced similar ideas, suggesting that the pressure of air resulted from hard, rapidly moving particles impinging on the walls of the enclosing vessel. He tried to deduce Boyle's Law—'as the volume of a gas is changed, the pressure changes proportionately with the density'—but failed. Sixty years later Daniel Bernoulli (1700–82), Professor in the University of Basle, showed that the law would be true if the particles were infinitesimal in

* *New Experiments Physico-Mechanicall touching the Spring of the Air and its Effects*, 1660.

† *Syntagma philosophicum* (*Opera* I, 1658).

size, and also examined how the law would be modified if the particles were of appreciable size.

The subject now lay quiescent for nearly a century, and then revived with abounding vigour, stimulated by the investigations of Herapath (1821), Joule (1848), Krönig (1856), Clausius (1857) and Maxwell (1859).*

Joule calculated how fast the molecules of a gas would have to move in order to produce the observed gas pressure by their impacts, and found that their speed in ordinary air would have to be about 500 metres per second—roughly the speed of a rifle bullet. In warmer air, the molecules must, of course, move more rapidly, in cooler air more slowly. In general the energy of motion per molecule is proportional to the temperature, measured from the absolute zero at which the energy of motion is nil.

A more rigorous treatment of this and other questions of the same sort was produced by J. R. E. Clausius, Professor of Physics at Bonn, in 1857. He began by making three simplifying assumptions; first, that the molecules of a gas all move at the same speed; second, that they exert no forces except when actually in collision; and third that they are infinitesimal in size. He then showed that the pressure of a gas would be equal to two-thirds of the energy of motion of all the molecules in unit volume of the gas. Boyle's Law was explained at once. For if the gas were allowed to spread through double the volume it originally occupied, the total energy of molecular motion would remain the same, so that the energy per unit volume would be halved. Hence the pressure also would be halved. Suppose, however, that the gas were heated without its volume being allowed to increase, then the energy of molecular motion would increase

* There was also an important paper which the English physicist J. J. Waterston submitted to the Royal Society in 1845. This contained a calculation of the pressure and temperature of a gas in terms of the speed and mass of its particles. But there were many errors in it and, partly for this reason but partly through a real miscarriage of justice, it was not published until 1892, when it was printed for its historical interest.

proportionately with the absolute temperature, so that the pressure would also be proportional to the absolute temperature. This is the famous law of Charles and Gay-Lussac, which was first published by Gay-Lussac (1778-1850), in 1802, although it had been discovered experimentally by Charles (1746–1823), the balloonist, in about 1787.

Clausius also proved that the molecules of different gases must move with different speeds, the speed in any gas being inversely proportional to the square root of its molecular weight—a law which Thomas Graham had discovered about 1830 from experiments on the diffusion of gases through porous substances. Clausius also showed that the law of Avogadro (p. 267) followed as a consequence.

The three simplifying assumptions of Clausius are all untenable. The assumption that the molecules of a gas all move with the same speed is impossible, because the molecules of a gas must crash into one another at frequent intervals, and would change their speeds at every collision, so that, even if the speeds were all equal at any one moment, they would soon cease to be so. In 1859 James Clerk Maxwell, then Professor of Physics at Aberdeen, but afterwards at Cambridge, took up this subject, and tried to find what the average speeds of the molecules would be if the disturbing effects of collisions were allowed for, and also how the speeds of individual molecules would be distributed about this average. The result he obtained, commonly known as Maxwell's Law, was reached by brilliant mathematical insight rather than by rigorous mathematical analysis. No one defends his proof to-day, but everyone agrees that it leads to the correct result. At a later date (1887) the Dutch physicist Hendrik Antoon Lorentz gave a rigorous proof, based on a method which the Viennese Professor Ludwig Boltzmann had introduced in 1868.

Maxwell's Law shows that the distribution of speeds is very similar to the distribution of errors when a number of marksmen all aim at the centre on a target. There must, of course, be some differences since a target is a two-dimensional structure, while the motion of molecules takes place in three dimensions, but this is about the only difference there is. We may picture all the molecules as aiming at a particular speed, and their departures from this, which now become the measure of their failures, are distributed according to the well-known law of trial and error. The exact knowledge contained in Maxwell's Law was an essential preliminary to further progress, if this was to be exact.

The problem had also been simplified by supposing that molecules exerted forces on one another only at the moments of actual collision. Maxwell discarded this simplification and assumed that molecules exert repulsive forces on one another even when they are not in contact; these forces were supposed to exist at all distances, but were insignificant except at very short distances. Maxwell supposed that their intensity fell off as the inverse fifth power of the distance, certain experiments having led him to think that this was the true law of molecular force. On this assumption, he studied various properties of gases, especially conduction of heat, diffusion, and internal friction or viscosity. His work has recently been extended in various directions and to other laws of force by a number of mathematicians, especially Chapman, Enskog, Lennard-Jones and Cowling.

It is also impossible to suppose that the molecules of a gas are infinitesimal in size. The three properties of a gas just mentioned are all a consequence of its molecules being of finite size, and their amount depends upon this size; smaller molecules travel farther without collisions, and so penetrate farther into adjacent layers of the gas. Thus these phenomena provide a means of determining the size of molecules. The average molecule is found to have a diameter of about a hundred millionth part of an inch. In ordinary air it travels about a 400,000th part of an inch between successive collisions, and it takes about an eight thousand millionth part of a second for each such free path. Maxwell's supposition that molecules repel at all distances is, however, inconsistent with the phenomena of capillary attraction and surface tension in fluids.

In 1873 the Dutch physicist van der Waals discarded this assumption, and supposed molecules in close proximity to exert attractive forces on one another. He investigated how a substance would behave under such conditions, and found that it would be able to exist in three distinct states, which he immediately identified with the liquid vapour, and gaseous states. His investigation also gave a convincing explanation of certain results which had been obtained by Andrews in 1869 and 1876. Up to then it had been supposed that any gas could be liquefied by sufficient compression. Andrews found that this was not so; each gas had its own 'critical temperature', and so long as a gas was above this no pressure, no matter how great, could liquefy it. These experiments, and Van der Waals's theory explaining them, have led to an elaborate technique for the liquefaction of gases and the various vast industrial applications of refrigeration. Before the end of the nineteenth century every common gas had been liquefied except helium. The Leyden physicist Kamerlingh Onnes liquefied this in 1908, the critical temperature being −268° C., or only 5° above the absolute zero of temperature.

ELECTRICITY

Important though all these advances were, the two centuries now under discussion were even more noteworthy for the vast and rapid development of the sciences of electricity and magnetism; they saw the inauguration of the electrical age, and the introduction of electricity into common life.

The foundations of the modern sciences had been laid in 1600, with Gilbert's publication of *De Magnete* (p. 152). This dealt mainly with magnetism, only one chapter being on electricity. The disproportion was natural at the time, for two reasons. First, magnetism had a practical value in the science of navigation. The mariner's compass—a small pivoted magnet—had been invented by the Chinese in the eleventh century, had been introduced into Europe by Mohammedan sailors,* and had been in common use ever since. Electricity had no such useful applications. Secondly, the phenomena of magnetism were fairly widely known, and were easily exhibited; anyone could carry a small lodestone in his pocket, while electrical effects were not so easily demonstrated.

Gilbert found the subject of magnetism encrusted in superstition; lodestones were credited with magic powers to cure diseases, but lost these if they were smeared with garlic; and so on. Gilbert ignored all this, and started on an experimental study of the attraction of lodestones, which led him to describe the earth as a huge magnet.

Electricity presented a somewhat different picture. It had been known from antiquity that when a piece of amber was rubbed in the proper way it acquired the power of attracting light objects, and Gilbert explained this by saying that electricity (the electric property; ἤλεκτρον = amber) resided in the rubbed object. Gilbert began by making himself a crude electroscope, which was simply a light metal needle which could turn horizontally on a pivot—a sort of electrical mariner's compass. This instrument could both show the presence of electricity, and give a rough indication of its amount. Gilbert rubbed a variety of substances, and examined their effects on this swinging needle. He found that the substances fell into two distinct classes. Some—such as glass, sulphur and resin—resembled amber in attracting the needle after they had been rubbed, but others—such as copper and silver—showed no such power. He described substances of the first type as 'electrics', and those of the second type as 'non-electrics'. Gilbert's non-electrics were the substances we now call conductors; electricity runs freely through them, so that they cannot retain a charge of elec-

* Sarton, *Introduction to the History of Science*, p. 756.

tricity for more than a minute fraction of a second. Gilbert's electrics were, of course, our insulators; electricity does not pass freely through these, so that it can be stored in them. It can also be stored in conductors if insulators are put so as to block its way of escape; otherwise trying to store electricity in conductors is like trying to store water in a sieve. But as Gilbert did not know this, he did not insulate his electroscope, and so could never observe the repulsion of electric charges. Cabaeus first observed this in 1629,* when he found that filings of metal which were attracted by rubbed amber would often jump several inches away from the amber after making contact with it. In 1672 Otto von Guericke, the inventor of the air-pump, exhibited the same phenomenon in a striking way.† He mounted a sphere of sulphur on an iron shaft, and turned this with one hand, while he rubbed it with the other, thus electrifying it somewhat vigorously. Small light bodies, such as feathers, would now jump about between the sulphur and the floor, as the electricity on the sphere alternately attracted and repelled them. A few years later Isaac Newton showed a similar experiment to the Royal Society, generating his electricity by rubbing glass with silk.

In 1733 C. F. du Fay tried to explain these attractions and repulsions on the supposition that all substances contained two kinds of electric fluid. These were usually present in equal quantities, and then they neutralised one another; for this reason he called them vitreous and resinous electricity. But they could be separated out by friction, and then they attracted or repelled one another according as they were of different or the same kind. The famous American, Benjamin Franklin (1706–90), then a printer in Philadelphia, suggested in 1747 that the attractions and repulsions were better explained by supposing that a single 'electric fire' or 'electric fluid' existed in all bodies. A body which had more than its

* *Philosophia magnetica*, II, 21.

† *Experimenta nova Magdeburgica de vacuo Spatio.*

normal share of the fluid was said to be plus, or positively, charged, while one with a subnormal share was minus or negatively charged. This one fluid explanation of electric phenomena prevailed for more than a century, but the older two-fluid explanation fits our present knowledge better.

Little further progress could be made without a better technique for the handling and storing of electricity. About 1731 it was found that electricity could be transferred from one body to another by putting a conductor to connect the two, and in 1745 it was found independently by P. van Musschenbroek, at Leyden, and by von Kleist at Kummin, that electricity could be stored in quantity in a 'condenser' formed by two conducting plates which had a slab of insulating material between them—the Leyden jar, named after the town in which it had been invented.

In 1749 Franklin suggested that lightning was an effect of electrical conduction, and confirmed this in 1752 by a painful experiment in which lightning was brought down from a kite flown in the sky to a key held in his hand. He proposed to protect buildings from the harmful effects of lightning by the use of 'lightning conductors'—metal rods which would conduct the lightning harmlessly into the earth.

In 1767 Priestley wrote a *History and Present State of Electricity*, in which he recorded and surveyed all the electrical knowledge of the time. Franklin had once dropped small balls of cork into a highly electrified metal cup, and found that they showed neither attractions nor repulsions. Priestley repeated this experiment in 1767, and further proved that there was no charge in the interior of an electrified conductor. If, for instance, a hollow conducting sphere has a small door made in its surface, it might have been expected that if the sphere were electrified, some of the charge could be conveyed out through the door. Actually this proves impossible because there is none inside to convey; the whole charge resides on the outer surface of the sphere. Comparing this with a mathematical theorem of Newton, which proved that there is no gravitational force in the interior of a hollow sphere, Priestley concluded that the law of force between electric charges must be the same law as for gravitational force, namely that of the inverse square of the distance. Cavendish obtained a direct experimental confirmation of this law in 1771, but his results were not published until 1879. In the meantime the French engineer Charles Augustus Coulomb (1736–1806) had again confirmed the law in 1785, by directly measuring the forces required to keep two small electrified pith balls at a series of measured distances apart.

On the basis of this result, Coulomb proceeded to build up a mathematical theory of electric force. His work was extended by a succession of still more distinguished theorists, including the great French mathematician Simeon Denis Poisson (1781–1840), and the equally eminent German mathematician Karl Friedrich Gauss (1777–1855).

THE ELECTRIC CURRENT. By the end of the eighteenth century the science of electric charges at rest—electrostatics, as we now call it—had attained pretty nearly to its present form; nothing of fundamental importance was added until the electron appeared nearly a century later. But new territory was being opened by the discovery of the electric current—electric charges in motion.

Like so many of the fundamental discoveries in physics, this was almost accidental. About 1773, physicists became interested in the shocks produced by the so-called 'electric' fishes, the *Torpedo* and the *Gymnotus*. As these shocks felt rather like those received from Leyden jars they were presumed to be electrical in origin, and this led to a study of the biological aspects of electricity. In 1786 the Italian Luigi Galvani of Bologna passed the discharge from a Leyden jar through the leg of a frog, and observed spasmodic contractions in the muscles. He also imagined that he had shown that electric phenomena could be produced by muscular contraction, which is true, although he had not proved it. He attributed these effects to what he called animal electricity, but other people called 'galvanism'.

In 1800, Alessandro Volta of Pavia (1745–1827) showed that electric activity could stimulate the organs of touch, taste and sight, and thus produce a variety of bodily sensations. For example, if two coins of different metals were placed, one above and one below the tongue, and their two surfaces were put in electric contact by a wire, the tongue experienced a salty taste. He attributed this to animal electricity, because he thought it was dependent on the presence of living matter —the tongue. But he soon found that electrical effects were still produced when the tongue was replaced by a thickness of cardboard steeped in brine. He then made a 'Volta pile', in which layers of zinc, paper, copper, zinc, paper, copper, ..., copper, were placed on top of one another to any number but always in this order, beginning with zinc and ending with copper. If the bottom zinc was now joined directly to the top copper, electricity was found to flow continuously through the wire. This simple piece of apparatus was the prototype of electric batteries and accumulators, and all electric current was produced by some such means as this until the dynamo came into use at a later date.

In 1801 Wollaston showed that electricity generated in this way produced exactly the same effects as the animal electricity of Galvani. The next year Erman used an electroscope to measure the degree of electrification produced by voltaic piles, and found that a pile with a great number of layers produced only the same degree of electrification as a very little friction.

At this time the electricity of the voltaic pile was usually described as electricity in motion, and that produced by friction as electricity in tension. Then, in 1827, Georg Simon Ohm (1787–1854) replaced these vague descriptions by a more exact and scientific terminology. Comparing the flow of electricity to a flow of water, the flow of electricity in a frictional machine is like that of a waterfall in which a small quantity of water falls from a great height, while the flow in the connecting wire of a voltaic pile is like the flow of a broad river in which a vast mass of water experiences but little fall. The continuous flow round a circuit is like the flow of water in a circular canal which falls everywhere except that at one point a pumping station raises the water back to its original level. The pumping station may be a voltaic pile or a battery or accumulator or even a dynamo. Ohm introduced precision into the ideas of quantity of electricity, current strength, and electromotive force. A quantity of electricity corresponds to a quantity of water in our analogy, and a current to the current in our canal—the flow past a given point in a unit of time—while the electromotive force (or voltage) represents the height through which the river falls; strictly it is defined to be the amount of work that must be done to carry a unit quantity of electricity round a circuit, or more generally from one point of it to another.

Electro-chemistry. In the same year 1800 in which Volta constructed his first pile, two Englishmen, Nicholson and Carlisle, made a modification of it which led to momentous consequences. They did not connect the first zinc directly to the last copper by a single wire, but joined a brass wire to each, and led the two loose ends into a vessel of salt and water, the salt having been added to turn the liquid into a conductor of electricity. Instead of passing through Volta's single wire, the current of electricity now passed through the two brass wires and the water, and as it did so, hydrogen gas was found to accumulate at the end of one wire, while the end of the other wire became oxidised. Or, if the wires were of a non-oxidisable substance, oxygen accumulated at the end of one wire and hydrogen at the end of the other. Clearly the passage of electricity had resolved the water into its two constituents, oxygen and hydrogen.

Out of this fundamental experiment, the whole subject of electro-chemistry has grown. Other substances besides water were soon resolved into their fundamental constituents in the same way, new elements even being discovered in the process; Sir Humphry Davy (1778–1829) discovered sodium and potassium in this way in 1807.

In the original experiment of Nicholson and Carlisle, the amounts of oxygen and hydrogen which were set free would be proportional to the amount of current which passed through the water. In 1833 Faraday (p. 284) measured how much electricity was needed to liberate a gram of various assigned substances by its flow, and found that the amount was closely related to the atomic weight of the substance. But it was even more closely related to the number of atoms that it liberated, and this, as Helmholtz subsequently remarked, pointed to some fundamental unit of electric charge being associated with every atom; it was the first foreshadowing of the electron.

It is so easy to measure the amount of substance liberated by an electric current that for a long time this was used for the practical measurement of electric current, and it is still used to define the unit of current. This unit, the 'ampere', is defined to be the current which liberates silver at the rate of 0·0011183 gram per second.

Faraday introduced most of the present-day terminology of electro-chemistry. The process of decomposing a substance into its simpler constituents by a flow of electricity he called 'electrolysis'—an electrical loosening—and the resolved substance he called an 'electrolyte'. The two plates plunged into the electrolyte were called the 'anode' and the 'cathode', the flow of positive electricity being from the former to the latter. The two constituents of a decomposed substance were called the 'anion' and the 'cation' respectively, or might be referred to collectively as 'ions'.

In this way electricity became linked up with chemistry, and it soon became associated with other branches of physics as well. The passage of electricity through a conductor was found to produce heat and sometimes also light, so that electricity became linked up with light and heat, and the foundations were laid for the modern technologies of electric heating and lighting. But a far more fruitful connection was soon to be established—that of electricity with magnetism.

Electromagnetism. Many observers had noticed that magnetised needles could be affected by electric discharges in their vicinity, and had concluded that there must be some link between electricity and magnetism. The first definite contact was established in 1820, when Professor H. C. Oersted of Copenhagen found that a magnetic needle which was balanced on a pivot or hung by a thread was deflected from its position when an electric current was flowing in its vicinity. Out of this single observation grew the whole technology of the electric telegraph; the tapping of a 'telegraph-key' at one point in an electric circuit alternately makes and breaks the current in the circuit, and so causes a succession of deflections of a magnetic needle at some other point in the circuit, possibly hundreds of miles away, and these can be interpreted according to a prearranged code. The same observation made it possible to measure the intensity of an electric current by the instrument called the 'galvanometer'. In the simplest form this consists of a magnetic needle which is suspended or pivoted so that it can turn freely, and is surrounded by a coil of wire which forms part of an electric circuit. The coil usually contains many turns of wire, so that even a small current passing through the coil will produce a measurable deflection of the needle, and the amount of the deflection gives a measure of the amount of current.

The French physicist André Marie Ampère saw at once that Oersted's discovery might have far-reaching consequences. It was common knowledge that two magnets *A* and *B* in proximity deflected one another, and Oersted had now shown that magnet *A* could be replaced by an electric current. Why should it not be possible to replace magnet *B* also by an electric current? In brief, a current seemed to be equivalent to a magnet; why should not two currents be equivalent to two magnets, in which case two currents would deflect one another as two magnets did? On 21 July 1820, just a week after the news of Oersted's discovery reached Paris, Ampère tried the experiment and obtained the result he had anticipated: two currents attracted or repelled one another just like two magnets.

This was a great achievement, but it formed only a stepping-stone to a far greater discovery by Michael Faraday, one of the greatest experimenters that science has ever claimed.

MICHAEL FARADAY (1791–1867) was born in Newington Butts, London, on 22 September 1791, being the son of a blacksmith and one of a family of ten children. He began life as an errand boy for a bookbinder to whom he was afterwards apprenticed. While he was thus employed, a customer gave him tickets for some scientific lectures which Sir Humphrey Davy was giving at the Royal Institution. Here he attracted the interest of Sir Humphrey, and finally was appointed his lecture assistant, thus escaping from the commerce he found distasteful to the science he loved.

He was, of course, familiar with the phenomenon of electric induction. When an electrified body, say a piece of amber, is brought near to an unelectrified body, the forces from the former body separate out the two kinds of electricity in the latter, the further parts becoming charged with electricity of the same sign as that on the amber, and the nearer parts with electricity of the opposite sign; this is why rubbed amber attracts light pieces of paper, or the needle of an electroscope. Faraday knew, too, that a magnet which is brought near to an unmagnetised piece of iron induces magnetism in the latter, so that it and the magnet attract one another; this is why a magnet picks up iron filings.

Faraday, pondering on these facts, wondered whether a current flowing in a circuit might not in the same way induce another current in a near-by circuit. With magnets, like induced like; why should it not be the same with currents which were so nearly equivalent to magnets? From about 1821 onwards, he tried, but without success, to induce a

Faraday's Christmas Lectures. 1855

The Prince Consort may be seen in the audience

current in a circuit by means of a magnet or a current in another circuit. Finally, in August 1831, he wound 203 feet of copper wire round a large block of wood, interspersing another 203 feet of similar wire as a spiral between its turns, and preventing electrical leakage between the two wires by binding both with twine. The ends of one wire were now joined to a galvanometer and the ends of the other to an electric battery of a hundred pairs of plates. This gave him two circuits in the closest proximity, with facilities for making a current in one and for observing a current in the other. If the flow of a current in the first circuit induced a current in the second, then the galvanometer included in the latter ought to show a deflection. At first, to his great disappointment, none could be observed. Finally, he noticed a slight instantaneous movement of the galvanometer in the second circuit, just at the moment of starting the current in the first circuit. A similar movement occurred in the reverse direction at the moment when the current was stopped.

The secret was now out. No current was induced in the second circuit by a steady current in the first, but one was induced by changes in this current. Once this had been established, it was only a step to see whether the current in the first circuit could be replaced by a magnet, and Faraday soon found that the motion of a magnet in the proximity of a circuit induced a current in the latter.

In this way the mechanical work of moving a magnet could be made to produce an electrical current. A way had been found to convert mechanical energy into the energy of an electric current. On a larger scale the magnet might be moved by a coal-burning engine, thus transforming the heat energy of coal into the energy of an electric current. The science of electrical engineering had been born.

For out of this fundamental experiment of Faraday has grown the whole vast technology of the mechanical production of electric power, and of the structure and operation of dynamos. The converse procedure of transforming the energy of an electric current into mechanical energy lies at the basis of electric motors and of all forms of electric transport—trains, trams, lifts and so on.

The same fundamental experiment also lay at the roots of the new mathematical theory of electro-magnetism which was shortly to be developed. In trying to explain his experiment Faraday adopted the same conception of lines of force as Gilbert had used in his *De Magnete*, supposing that magnets or electric charges or an electric current produced forces in the ether which was imagined to fill all space. If a thin card is laid on top of a magnet, and iron filings are scattered on the card, the filings are found to group themselves into chains which run approximately from one pole of the magnet to the other. Faraday imagined that these not only showed the presence of magnetic forces in the ether, but also mapped out the directions in which these forces acted at the various points on the card. He conjectured that there would be similar lines of electric force in the proximity of electric charges. Thus the ether itself would be in a state of pressure or tension, its pushes and pulls of course explaining the apparent attractions or repulsions of magnets and of electrically charged bodies. Faraday could now explain his fundamental experiment by supposing that a moving magnet carried lines of magnetic force about with it, and that when these swept across a conducting circuit, they caused a current of electricity to flow through the circuit.

Faraday had not intended his conception of lines of force to be interpreted very realistically, but only as a picture to help the mind grasp the mode of action of electric and magnetic forces. The whole train of ideas would obviously have been better expressed in mathematical terms, but Faraday was no mathematician. Clerk Maxwell, on the other hand, was, and in 1856 he published a paper 'On Faraday's Lines of Force', in which he tried to express Faraday's ideas in exact mathematical language. He followed Faraday in attributing electric and magnetic action to pressures and tensions in the ether, and showed in 1864 that any disturbance created in the ether by electric or magnetic changes would be propagated through it in the form of waves. In such waves the electric and magnetic forces would be at right angles to one another, and also to the direction in which the wave was travelling.

These waves would be propagated at a uniform speed which could be calculated, and proved, to within the limits of experimental error, to be precisely equal to the speed of light. In Britain, at least, this was generally held to establish that light was an electro-magnetic phenomenon, consisting of waves of electric and magnetic force travelling through the ether. The question 'What is light?' which had stood unsolved through the ages could be answered at last; light was a passage of electric and magnetic forces through the ether.

The continent, however, remained comparatively unconvinced. In 1879 the Berlin Academy offered a prize for a subject which had some bearing on the question. Helmholtz, who was then Professor of Physics at Berlin, brought this to the notice of Heinrich Hertz (1857–94), who was then one of his students but was later Professor at Karlsruhe and Bonn. The result was that in 1887 Hertz succeeded in making electrical sources in his laboratory emit waves which were of precisely the kind predicted by Maxwell's theory, and were shown to possess all the properties which Maxwell's theory required of them, as well as the known properties of light-waves, except that they were of much greater wave-length. They were in fact what we should now describe as radio-waves of very short wave-length. Out of these investigations of Faraday and Maxwell and the pioneer experiments of Hertz, the present vast technique of radio-transmission has grown.

After this experiment of Hertz, the continent followed Britain in accepting Maxwell's electro-magnetic theory of light.

Hertz further showed that Maxwell's electro-magnetic

laws implied an essential symmetry between electric and magnetic action. A change in the electric field of force set up magnetic forces; a change in the magnetic field of force set up electric forces; and the mathematical laws connecting the change with the new forces were the same in the two cases—a discovery which was of very great importance for later discussions as to the ultimate meaning of electricity and magnetism.

CHAPTER VIII

THE ERA OF MODERN PHYSICS (1887–1946)

THE two centuries from 1687 to 1887 may appropriately be described as the mechanical age of physics. Science seemed to have found that we lived in a mechanical world, a world of particles which moved as the forces from other particles made them move, a world in which the future is completely determined by the past. In 1687 Newton's *Principia* had interpreted the astronomical universe very successfully in this way. Before 1887 Maxwell had interpreted radiation in an essentially similar way, teaching that it consisted of disturbances travelling through an ether under the direction of mechanical laws. Finally, in 1887 Hertz produced radiation of Maxwellian type from electric sources in the laboratory, and demonstrated its similarity to ordinary light. This seemed to fit a final keystone into the structure which had been built up in the preceding two centuries.

Most physicists now thought of this structure as standing foursquare, complete and unshakable. It was hard to imagine the physicists of the future finding any more exciting occupation than dotting the i's and crossing the t's of the mechanical explanation of the universe, and carrying the measurement of physical quantities to further decimal places.

Little did anyone imagine how completely different the actual course of events would be. Yet the year 1887, which had provided a keystone to the structure, also saw the structure begin visibly to totter; it was the year of the famous Michelson-Morley experiment, which first showed that there was something wrong with the foundations. It was, as we can now see, the culmination of the mechanical age in physics and the inauguration of a non-mechanical era.

This does not mean that science had been following an entirely wrong road for two centuries. It had at least discovered a system of laws which gave a perfect, or almost perfect, description of the motion of planets and projectiles, of falling stones and rolling spheres; it had shown that moderate-sized objects in general behaved as though nature was completely mechanical. This was all good solid progress. But science was now beginning to investigate nature under a much wider range of conditions. A study of the very great and the very small might conceivably have shown that a mechanical picture was still adequate, even in regions which were very remote from direct human experience; actually it showed the reverse—the picture as a whole was in need of radical amendment. The history of physics since 1887 consists in large part of the story of this amendment.

ABSOLUTE SPACE. The first great amendment was the erasure from the picture of the absolute space which Newton had introduced as its framework. The ether, which formed an unobtrusive background to the whole picture, was supposed to serve a double purpose; it provided a fixed framework against which distances in space could be measured, and it transmitted radiation in the form of electromagnetic waves. But there was no experimental proof of its existence; this was purely hypothetical. The Michelson-Morley experiment was designed to get on closer terms with this elusive ether, in particular by measuring the speed at which the earth moved through it.

Light was supposed to travel through it at a uniform speed of roughly 186,300 miles per second—the velocity of light. But this speed might seem different to an observer who was stationed on a moving earth. If this moved through the ether in the same direction as the light, at a speed of x miles per second, then after one second the light would have travelled 186,300 miles *through the ether*, but as the earth would itself have moved through x miles of this, the light would be only $186,300-x$ miles *in front of the earth*. Thus the apparent speed of motion of the light—the true speed relative to the earth—would be only $186{,}300-x$ miles per second. If the light moved in the opposite direction to the earth, its relative speed would be $186{,}300+x$ miles per second. Now imagine a beam of light being sent out from a terrestrial source, moving in the same direction as the earth until it fell on a mirror, and then being reflected back to its source. It would perform the outward journey at $186{,}300-x$ miles per second, and the return journey at $186{,}300+x$ miles per second. Simple arithmetic shows that the double journey would take slightly longer than if the earth stood at rest in the ether,* and that the faster the earth moved, the greater this loss of time would be. Thus from the observed amount of lost time it ought to be possible in principle to determine the speed x of the earth's motion.

THE MICHELSON-MORLEY EXPERIMENT. To perform the experiment in the simple form just described is of course completely impossible; it would need stop-watches of incredible precision. But in 1887, the two American Professors Michelson and Morley devised a variant of it which seemed practicable, and likely to give the information wanted. They arranged for a beam of light to be divided into two halves, one-half being made to perform a double journey of the kind just described, while the other, acting as a sort of control, performed a double journey of the same length in a direction at right angles. When the two half-beams returned to the source, they were reunited and passed through a small telescope.

If the earth were standing at rest in the ether, then the two beams would of course take just the same times to perform their journeys; if they were started off together, they would

* The time of a double journey of length l would of course be

$$\frac{l}{186{,}300-x}+\frac{l}{186{,}300+x},$$

which is greater than $\frac{2l}{186{,}300}$ by an amount $\frac{2l}{186{,}300}\times\frac{x^2}{(186{,}300)^2-x^2}$.

arrive back together. But if the earth were in motion, the times of their journeys would be slightly different, and the difference could be made to show itself by interference (p. 250). The amount of the observed difference ought then to disclose the speed of the earth's motion. The method was so sensitive that a speed of less than a kilometre per second ought to have produced an observable result.

With such hopes the experiment was planned and performed. But no difference of times could be detected; everything seemed to happen just as it would if the earth were standing at rest in the ether. Of course the earth might have been standing at rest at the moment of the experiment, its 19 miles per second motion round the sun being just neutralised by a 19 miles per second motion of the sun in the opposite direction through space. If so, it was only necessary to wait six months, and the earth would then be moving through space at a speed of 38 miles per second. But precisely the same result was obtained when the experiment was repeated six months later, and on many subsequent occasions; the earth seemed always to stand at rest in the ether. It might have been thought that the earth was dragging the ether along with it, had not such a possibility been precluded by the phenomenon of aberration (p. 241); this definitely required the earth to move freely through the ether.

For a time the whole situation seemed to be one of complete mystery. The situation cleared when the same solution was proposed independently and almost simultaneously by Lorentz of Haarlem (p. 273) and George Francis Fitzgerald (1851–1901) of Dublin. The two half-beams had taken the same time to perform their journeys, although their average speeds had been different, and the inference seemed to be that the journeys must have been of different lengths. The whole situation could be explained by supposing that the motion of an object caused it to shrink in the direction of its motion, but not in a perpendicular direction, by just enough to compensate for the difference in the speeds of the two half-beams of light.* Such a shrinkage could never be detected by direct measurement, since every yardstick would shrink just as much as the object it was measuring. But Lorentz showed that a shrinkage of precisely the required amount was actually predicted by Maxwell's electromagnetic theory, so that the Michelson-Morley experiment could have given no other result than it did; in fact it amounted to nothing more than a confirmation of this theory.

Yet if there were an ether, the earth must move through it, and it seemed inconceivable that the motion could cover up its tracks so completely as to elude all the resources of experimental science. Nevertheless, the inconceivable happened; a great number of other experiments were devised to discover the earth's motion through the ether, and all gave the same result: If there was an ether, things were the same as they would be if the earth stood permanently at rest in it.

The Theory of Relativity

In 1905 a new turn was given to the whole discussion by Albert Einstein, who was then an examiner of patents in the Patent Office at Berne. He had been born of Jewish parents at Ulm on 14 May 1879, educated at Munich, in Italy and at Aarau in Switzerland, and had taught in Zürich and Schaffhausen before going to Berne. He subsequently held Professorships or similar posts in Zürich, Prague, Zürich again, the Kaiser Wilhelm Physical Institute in Berlin, Oxford and Princeton.

We have already noticed instances of scientific results being deduced from principles of a very general kind, such as the impossibility of perpetual motion which had been employed in this way by Leonardo (p. 124), by Stevinus (p. 144), and by Helmholtz (p. 270). Einstein thought that the accumulation of experimental results just mentioned

* For the compensation to be exact, the shrinkage would have to be in the ratio of $\sqrt{(1 - v^2/c^2)}$ to 1, where v is the speed of motion, and c is the speed of light.

might point to the existence of a similar general principle, which he enunciated in the form: 'It is impossible to determine the speed of motion of an object through space by any experiment whatever.' Experience had shown that it was impossible in practice; Einstein now suggested that it was also impossible in principle—not because human skill was inadequate to find the way, but because the constitution of the world, the laws of nature, made it impossible.

The new principle clearly implied that all the phenomena of nature must be the same for a person moving with one speed as for a person moving with some quite other speed. This at once explained the negative results of the Michelson-Morley and of all similar experiments. It further showed that nature was not concerned with absolute speeds, but only with relative speeds. Hence it was called the 'Principle of Relativity'.

It suggested, and even encouraged, a new interpretation of physical phenomena, and a new view of the aims of science. The speed of light, it now appeared, was not constant relative to an absolute space pegged out by a material ether, but relative to an observer; he, and not the ether, now became the central fact of the situation. As the subject developed, it became clear that the phenomena of nature were determined by us and our experiences rather than by a mechanical universe outside us and independent of us. Democritus, back in the early infancy of science, had taken the emphasis off us and our sensations, and had transferred it to an objective nature outside us; the new principle now took it off this, and retransferred it to us and our subjective measurements.

If the new principle was a necessary consequence of the constitution of the world, it ought to tell us something about this constitution. Einstein accordingly set to work to examine the physical implications of the new principle. He found that it was entirely consistent with Maxwell's electromagnetic theory, but was in some ways inconsistent with the

Newtonian mechanics. For instance, the mass of an object would now depend on its speed of motion. J. J. Thomson had already shown that this was true for a body charged with electricity; the new principle required it to be true for every moving object. The principle further required all energy to be possessed of mass; for instance, the kinetic energy of motion of a rapidly moving object endowed it with the extra mass just mentioned. So also, any object which was losing energy must be losing mass also; the sun's radiation, for example, implied a diminution of its mass at the rate of 250 million tons per minute, and the problem of finding the source of the sun's radiation involved that of finding how the sun could lose so much mass. In brief, the two great principles of Conservation of Mass and Conservation of Energy became identical.

SPACE-TIME. In 1908 the Polish mathematician Minkowski stated the whole content of the theory in a new and very elegant form. Hitherto the laws of nature had been thought of as describing phenomena which occurred in three-dimensional space, while time flowed on uniformly and imperturbably in another and quite distinct dimension of its own. Minkowski now supposed that this fourth dimension of time was not detached from and independent of the three dimensions of space. He introduced a new four-dimensional space to which ordinary space contributed three dimensions, and time one; we may call it 'space-time'. Every point in space-time resided in three dimensions of ordinary space and in one dimension of time, and so would represent the position of a particle in ordinary space at an assigned instant of time. The succession of positions which a particle occupied in ordinary space at a succession of instants of time would be represented by a line in space-time; this he called the 'world-line' of the particle.

When we formulate the laws of ordinary optics, we usually think of light as travelling in three dimensions. Gravity dominates our everyday lives so much that we almost instinctively think of ordinary space as consisting of two horizontal dimensions and one vertical. But ordinary optics knows nothing of gravity, with the result that the laws of optics recognise no distinction between vertical and horizontal. Space may be divided into its three dimensions in some quite other way, but the laws of optics will remain formally the same as they were before. Now just as the laws of optics refer to a three-dimensional space in which no distinction is made between horizontal and vertical, so the laws of nature refer to a four-dimensional space-time in which no distinction is made between space and time—such, at least, is the content of the theory of relativity as transformed by Minkowski. In other words, nature refuses to break up the four-dimensional space-time into an absolute space and an absolute time for us.

In this way Newton's absolute space and absolute time fell out of science, and they carried much with them in their fall. First to go was the concept of simultaneity. If there is no gravity, vertical height means nothing, and it is meaningless to say that two points are at the same height. In the same way, if there is no absolute time, it is meaningless to say that two happenings at different points occurred at the same instant, or even, except in certain cases, that one happened before the other. This made Newton's law of gravitation become meaningless. It means nothing to say that a sun at S attracts an earth at E with a force which depends on the distance ES, unless we can attach some definite meaning to ES. To do this we have to know the position of the sun at one instant, and that of the earth *at the same instant*, and if simultaneity means nothing, then this can mean nothing. It now became necessary to find some way of treating gravitation which should not involve simultaneity. Einstein found this through the medium of his 'Principle of Equivalence'.

THE PRINCIPLE OF EQUIVALENCE. When the aeroplane in which we are travelling makes a rapid turn, our ideas of horizontal and vertical are apt to become strangely confused;

gravity seems to have changed its direction. The explanation is that the plane and all its contents are experiencing an acceleration which produces effects precisely similar to those of gravitation—so much so that the acceleration and the original force of gravitation seem to combine into a new force of gravity which acts in a new direction. In the same way, when a lift is rapidly started or stopped, the acceleration acts like a new force of gravitation and combines with the old force so indistinguishably that our bodily weight seems to undergo a sudden change.

Such considerations led Einstein in 1915 to propose a new principle—the 'principle of equivalence'—to the effect that gravitation and the pseudo-gravitation of an acceleration not only seem to be similar, but actually are so similar that no experiment can distinguish between them at any single point. This was found to involve surprising consequences. We have seen how any division of Minkowski's space-time into ordinary space and time is a matter of our own choice and so is subjective. To state the laws of nature objectively, we must not state them in terms of ordinary space and time, but in terms of Minkowski's space-time as a whole; that they can be so stated is the whole content of the principle of relativity. For instance, Newton's first law can be stated in the form that the world-line of a particle on which no forces act is a straight line. If, then, a particle is acted on by gravitation, its world-line cannot be a straight line. We might expect it to be a curved line, as, indeed, it proves to be, but the curvature is inherent in the Minkowski space-time, and not in the world-line. To give an objective description of the effects which had hitherto been attributed to a force of gravitation, Einstein found it necessary to think of space-time as curved. The requisite curvature may be compared to that of a rubber balloon, the four-dimensional space-time corresponding to the rubber surface, and not to the space inside or outside the balloon. There is a special curvature in the proximity of matter, although we must not say that the matter is either the cause or the effect of the curvature. In this curved space-time, the world-line of a particle, whether acted on by gravitation or not, is a 'geodesic'—a term which needs explaining.

Geodesics are, in brief, shortest distances. The term is used in geography (usually in the form 'geodetic lines') to denote the shortest path from place to place over the surface of the earth—the course of an aeroplane flying by the shortest route from airport to airport, or the course marked out by a string stretched tightly over the surface of a geographical globe. In mathematics the term is used more generally (usually in the form 'geodesic') to denote the shortest line from point to point either on a curved surface or through a curved space. On any uncurved surface or in any uncurved space the geodesics are of course straight lines.

Any region of space-time which has no gravitating mass in its vicinity is uncurved, so that the geodesics here are straight lines, which means that particles move in straight courses at uniform speeds (Newton's first law). But the world-lines of planets, comets and terrestrial projectiles are geodesics in a region of space-time which is curved by the proximity of the sun or earth, and so are not straight lines. No force of gravitation is, however, needed to impress curvature on the world-lines; the curvature is inherent in the space, just as the curvature in an aeroplane course is inherent in the curvature of the earth's surface.

The theories of Newton and of Einstein predicted the same motion for a body which was free from the action of all forces, namely, uniform motion in a straight line. They also predicted the same motion for a body which moved slowly under the action of a gravitating mass. So far, then, it was impossible to call on observation to decide between the two theories. But the two theories predicted different motions for bodies moving at high speeds, so that observations on such bodies might be expected to decide between the two theories.

According to Newton's laws, a planet must move round the sun in a perfect ellipse; according to Einstein's theory, this ellipse must rotate slowly in its own plane. It had long been known that the orbit of Mercury—the fastest moving of the planets—exhibits a motion of rotation of just this kind; Leverrier had first noticed it, and the American astronomer Simon Newcomb had measured its amount. Many unsuccessful attempts had been made to explain it; Einstein's theory not only explained it at once, but predicted its true amount.

Other observational tests were possible, for the principle of relativity required a ray of light to be deflected whenever it passed near to a gravitating mass. Thus stars which appear to lie near to the sun in the sky ought to be seen displaced from their proper positions. Such displacements could be observed only at a total eclipse of the sun, and they had not been observed when Einstein's theory appeared. They were found at once when the Observatories of Greenwich and Cambridge sent out expeditions to look for them at the eclipse of 1919, and their discovery immediately converted the scientific public to Einstein; since then it has been generally accepted that his scheme must supersede that of Newton.

Lastly, the principle of relativity requires that when light is generated in a region where gravitational forces act, as at the surface of a star, the lines in the spectrum of the light must be displaced towards the red end of the spectrum. The effect is so small that it is usually difficult to observe, but it is found unmistakably in the light emitted by very dense stars, where it is greatest in amount. Here it has become one of the ordinary working tools of astronomy, and provides a means of measuring the masses and diameters of dense stars.

This scheme of Einstein was in direct line of descent from Hero's principle (p. 81) that in ordinary space a ray of light takes the shortest path between any two points on it—in brief, that its world-line is a geodesic. So far as light is concerned, Einstein's scheme merely extends this principle to a curved space-time, and asserts that even near to gravitating masses the world-line of a ray of light is a longest (not a shortest) distance, although it can no longer be a straight line. It is much the same with the motion of material masses. The principle of least action determined the path of a planet by making a certain quantity a minimum; the new principle of Einstein also determines it by making a certain quantity a maximum, namely, the length of a world-line. It is of interest to compare this last solution with earlier efforts to explain the planetary orbits. We pass in review the complicated interconnected spheres of Eudoxus and Callippus, the intricate circles and epicycles of Ptolemy and the medievals, the ellipses of Kepler, Newton and most of the moderns, until finally we end up with the exceedingly simple 'longest distance' of Einstein. It is a striking instance of nature proving to be exceedingly simple when we approach her problems from the right direction.

The two schemes of Einstein and Newton are poles asunder in their physical interpretations, but it would be a mistake to think of the Newtonian scheme as nothing but an accumulation of errors. The quantitative error in Newton's law of gravitation is so small that nearly 200 years elapsed before any error was discovered, or even suspected. Indeed, the difference between the laws of Newton and Einstein depends on the square of the small ratio v/c, where v is the speed of a moving planet and c is the speed of light. Even for the fastest moving planet, Mercury, this square is only 0·00000003; throughout the whole solar system the two theories differ quantitatively only by this minute fraction or less. And when we come down from the heavens to the earth, we find a science of everyday life which is still entirely Newtonian; the engineer who is building a bridge or a ship or a locomotive does precisely what he would have done if Einstein's challenge to Newton had never appeared, and so does the computer who is preparing the *Nautical Almanac*, and the astronomer who is discussing the general motion of the planets.

After the theory had explained the apparent force of gravitation in such a simple way, it was natural that forces of other kinds should come under scrutiny, particularly those of electricity and magnetism. Faraday and Maxwell had thought that the existence of electric force at a point indicated a disturbance in, or a displacement of, the ether at that point, in which case the electric force would have a definite value at every point of space. The theory of relativity now showed that the force at a point was merely a matter of measurement. Different ways of measuring force gave different values, and all were equally right; absolute force was seen to be as illusory as absolute space and time. In the same way, the mass of a moving body was found to depend on its measured speed of motion, and this in turn depended on the measurer, or rather on the coordinate system he adopted. Thus absolute mass fell out of science, and as energy was proportional to mass, absolute energy fell with it. The idea that energy was localised in the various parts of space had to be abandoned.

Einstein and others have tried to construct a 'unitary field-theory' which should interpret both electromagnetic and gravitational forces in one comprehensive scheme as consequences of the properties of space. It would be wrong to say that no progress has been made along these lines, but it is certain that no complete success has been achieved, and at the moment there is no unitary theory which commands any large measure of assent. We are here at the very frontier of knowledge, and the interpretation of electromagnetic force lies in the unconquered unknown.

EXPERIMENTAL PHYSICS

While mathematical physics was obtaining these sensational results, experimental physics was experiencing its own series of sensations, which inaugurated the era of modern physics. The second half of the nineteenth century had produced great technical improvements in many of the instruments of physical research, and out of these a new physics was born. In particular, the air-pump was improved almost beyond recognition, so that every experimenter had high vacua at his disposal, and could study the phenomena of gases in which the free path was many millimetres in length. This led to the discovery of a minute electrified particle—the electron—which proved to be an ingredient of every kind of atom and, later, to the discovery that all matter was wholly electrical in its structure.

The Electrical Structure of Matter

When an electric current is turned off, a spark is often seen momentarily bridging the gap between the two terminals of the switch; the momentum of the current has carried it on in its old path for a moment, but is soon checked by the high resistance of the air in the gap—for air under ordinary conditions is a bad conductor of electricity. If the switch were placed in a vessel from which the air had been partially exhausted, the same phenomenon would be observed in a heightened form, because a rarefied gas offers less resistance to the passage of the current. With a sufficiently low gas pressure and a sufficiently high voltage, a current can be maintained permanently between two terminals at a considerable distance apart, and is found to produce a characteristic phosphorescence on those walls of a glass vessel on which it falls.

The observation that electricity passes more easily through a partial vacuum than through air at atmospheric pressure was first made by Watson in 1752. The phosphorescent glow was first recorded by Faraday in 1838. From 1859 on, the passage of electric currents through gases was studied in detail in Germany, especially by Geissler, Plücker and Hittorf. Hittorf showed in 1869, and Goldstein in 1876, that the current normally flowed in straight lines, and that its flow could be obstructed by placing a solid obstacle in its path. This was found to project a 'shadow' on the walls of the vessel, which showed itself by an absence of the characteristic phosphorescence. But only one of the two terminals, the cathode, threw such a shadow, which seemed to show that the electric current consisted only of one-way traffic—of negative electricity travelling from the cathode to the anode; for this reason, it was described as consisting of cathode rays.

In Germany the rays were thought to consist of waves, in which case the shadow thrown by an obstacle would be exactly like that thrown by a man standing in the sunshine. But when Varley and Crookes passed the rays between the poles of a powerful magnet, they found that the shadow changed its position; the rays had been bent out of their courses by the magnetic forces. Now it is a well-known fact of electromagnetism that magnetic forces deflect moving electrified particles from their courses, but do not deflect electromagnetic waves. Thus the observation of Varley and Crookes suggested that the cathode rays were showers of electrified particles which transported electricity just as a shower of rain drops transports water. In 1895 Perrin showed that a conductor becomes negatively charged when cathode rays fall on it, just as the pavement becomes wet in a shower of rain.

THE CATHODE RAYS. The problem of the nature of the cathode rays was solved in 1897, with epoch-making results. We have seen that they are charged particles. The amount by which a charged particle is deflected in a magnetic field depends on both the electric charge on the particle and on the mass of the particle; if the charge is great, the deflection is great, because the magnetic forces get a better grip on the particle, while if the mass is great the deflection is small, because there is more inertia resisting change. The amount of deflection depends also on the speed with which the charged particle is moving. If this speed of motion is known, a measurement of the deflection makes it possible to evaluate the ratio of the charge of a particle to its mass—a ratio which is commonly denoted by e/m. This was first done for the cathode-ray particles in 1890, by Arthur Schuster, then Professor of Physics at Manchester. Physicists were already familiar with one ratio of charge to mass, namely, that for the hydrogen ion in electrolysis. Schuster estimated that the ratio for the cathode particles was about 500 times greater* and, as he assumed that there could be no particles smaller than atoms, he concluded that the cathode particles must be highly charged atoms. But in 1892 Hertz found that the rays could penetrate thin metal foils to a degree which seemed impossible for particles of atomic size, so that if the rays consisted of particles, these must be of less than atomic size.

The subject was clearly of great importance, and many determinations were now made of the speed with which the particles travelled, and hence of the ratio of charge to mass. In 1897, J. J. Thomson in Cambridge and Wiechert in Germany independently arranged for their particles to be deflected by electric and magnetic forces simultaneously, so that the speed could be measured directly, and the ratio e/m deduced. They and many other experimenters found that this ratio was about 1800 times as great as for the hydrogen atom.

In 1896 a new piece of apparatus had been devised, which now proved to be of the utmost service. C. T. R. Wilson, afterwards Professor of Natural Philosophy at Cambridge, invented his famous 'Condensation chamber', in which electrically charged particles can be made to collect drops of water round them, just as they do in the atmosphere to form rain drops. An artificial shower of rain can be produced by letting these drops of rain fall on the floor of the chamber. The average size of the drops can be estimated by noting the speed at which they fall against the resistance of the air, and the total number in a shower can then be calculated by weighing the total fall of water. A measurement of the charge on all the water by an electroscope now makes it possible to evaluate the charge on each drop.

Using this piece of apparatus, Thomson found in 1899 that

* This is now known to have been an underestimate, the true value being about 1840.

the cathode particles carry the same charge as the hydrogen ions in electrolysis. He obtained the same value for particles obtained from quite other sources, as, for instance, from a plate of zinc under bombardment by ultra-violet radiation, while J. S. Townsend, afterwards Professor of Physics at Oxford, obtained similar results from a study of the rate at which gaseous ions diffused into one another. It was now clear that the high value of e/m for the cathode particles did not result from a high charge, but from a small mass; this could only be about $\frac{1}{1800}$ times the mass of the hydrogen atom.

All this persuaded physicists that they were dealing with particles far smaller than the hydrogen atom, which had hitherto been supposed to be the smallest particle in nature, and all matter now appeared to contain these new particles in its structure. They were found everywhere and were always the same, whatever the source from which they came; they were called 'electrons', a name which had been proposed by Johnstone Stoney of Dublin in another connection.

The Dutch physicist Hendrik Antoon Lorentz now extended Maxwell's electrical theory to cover the new facts. Light, for instance, would be produced by the motion of electrons inside the atom. The motion would, of course, be modified by the presence of magnetic forces, so that if a substance were to emit light in a magnetic field, this light would be different from that emitted under normal conditions. Now P. P. Zeeman, Professor of Physics at Amsterdam, had already observed an effect of this kind in 1896 (the 'Zeeman effect'). At first the modification seemed to consist of a mere broadening of the spectral lines, but when a more powerful magnet was employed, each line was seen to be split into a number of separate components. Lorentz now showed how this could be explained mathematically by supposing that the light was produced by the motion of electric particles in the atom, each carrying charges and masses equal to those of the electron.

It seemed safe to suppose that light was caused by the motion of electrons in the atom. Further, as the total charge on the normal atom was *nil*, it was plausible to assume that the normal atom consisted of a number of moving electrons together with just sufficient positive charge to neutralise the total charge on all the electrons.

Newton had supposed that the mass of an object remained the same through all changes of its motion, but as far back as 1881, J. J. Thomson had shown that this would not be the case if the object was electrified; Maxwell's equations required the mass of an electrified body to increase as its speed of motion increased, the reason being, in brief, that a charged body was not self-contained, but consisted in part of lines of force which stretched away from the body into infinite space. As the speed of the body increased, these lines of force rearranged themselves in a way which produced an increased opposition to further change; in a word, the apparent mass of the body increased. The mass of a body might thus be divided into two parts, an internal or Newtonian mass, which experienced no change, and an external or electrical mass, which depended on the speed of motion.

From about 1906 on, a number of physicists—Kaufmann, Bucherer, Bestelmeyer and others—investigated experimentally how the mass of a moving electron depended on its speed of motion, and obtained the sensational result that the dependence was exactly that which Thomson had calculated for the electrical part alone. In other words, the electron had no Newtonian mass, and its mass appeared to be wholly electrical. It was a reasonable assumption that the same would be true of the positive electricity of the atom, so that all matter appeared to consist simply of electricity. To the question 'What is the ultimate substance of the Universe?' which had puzzled science from the time of Thales on, it at last appeared possible to give an answer—the one word 'electricity'.

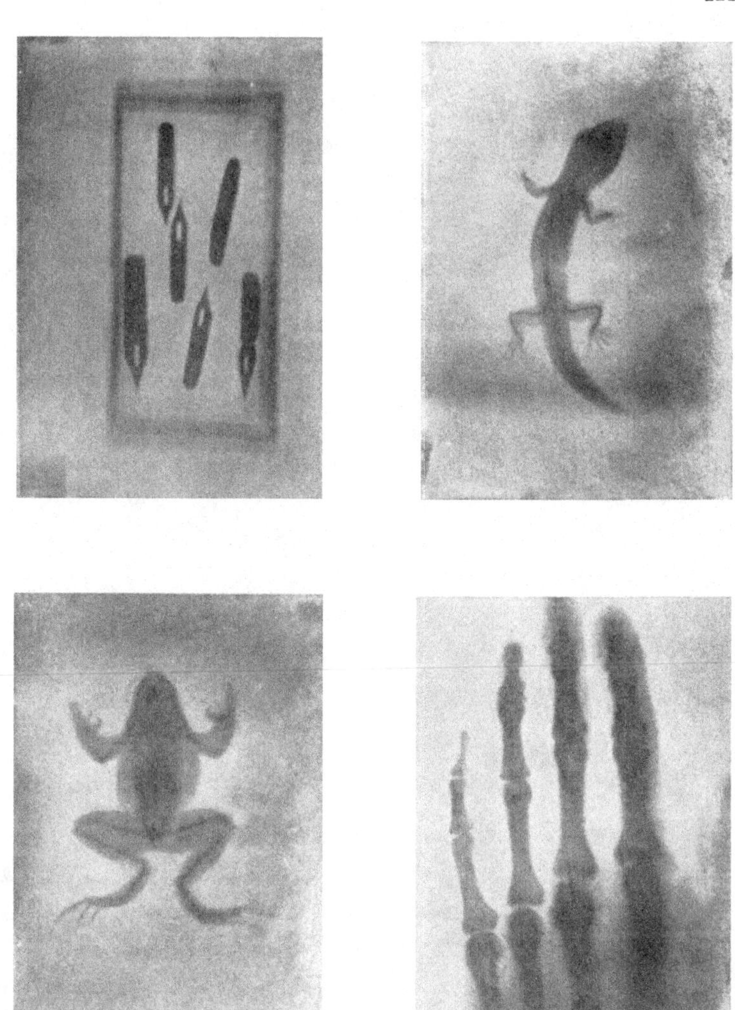

X-ray photographs taken by Hayles in January 1896

Showing pens in a box, a lizard, a frog, and a hand. They are believed to be the first X-ray photographs taken in England

X-RADIATION. The year 1895 is noteworthy for the discovery of the Röntgen rays, or X-rays as they are often called, by Wilhelm Konrad Röntgen (1845–1923) of Würzburg. Many experimenters had found to their sorrow that photographic plates which were stored near to electrical discharge tubes suffered damage through fogging, but most of them regarded this as a nuisance calling for a new place of storage rather than as a scientific fact calling for study. Röntgen, however, felt curiosity as to the cause of the fogging, and conjectured that the tubes must emit some hitherto unknown form of radiation which could pass through the material in which the plates were packed. Almost by accident he found that there was such a radiation; it was invisible to the eye, but could be detected from its property of making phosphorescent material become luminous. This property of the radiation made its study easy. It was soon found that a thick sheet of metal stopped it entirely, but that it could pass through a thin sheet of metal, or through light substances such as cardboard, paper, wood or human flesh, and could still affect a phosphorescent screen or a photographic plate after its passage. Thus it was possible to photograph the bones inside a living body through their covering of flesh, a discovery which interested the layman almost as much as the professional physicist, and proved to be of the utmost value to the surgical and medical sciences.

For many years physicists could not decide whether this radiation consisted of particles or of waves. If it consisted of particles, these must be uncharged, for the radiation was not deflected by magnetic forces. If it consisted of electromagnetic waves, these must be of very short wave-length, for the radiation was far more penetrating than visible light.

The problem was solved by three German physicists in 1912. Laue saw that if the radiation consisted of short waves, then the regularly arranged atoms of a crystal ought to diffract it, just as the regular scratches on a diffraction grating (p. 204) diffract visible light, and he calculated the general type of diffraction pattern which was to be expected on this hypothesis. When Friedrich and Knipping tested the suggestion experimentally, they found exactly the type of pattern which Laue had predicted.

This established that the radiation was of an electromagnetic nature, but it had wider consequences than this. Different arrangements of the atoms in a crystal would of course result in different diffraction patterns, so that the atomic arrangement could be deduced from the observed diffraction pattern. The new technique was developed rapidly by William Bragg and his son Lawrence Bragg, as well as by many others. The Braggs first studied very simple substances, such as the chlorides of sodium and potassium, and found that their atoms formed regular cubical patterns in which an atom lay at each corner of every cube. These, like other inorganic compounds they studied later, showed no pairing of the atoms into molecules; in the solid state, the atom had become the unit. But when Sir William Bragg investigated various organic compounds such as naphthalene and anthracene in 1921, he found that the molecules retained their identities as close clusters of atoms.

The X-ray analysis of solid substances has proved of the utmost value to metallurgy and biochemistry, but probably experimental physics benefited most of all from the discovery of X-radiation. For the radiation changed any gas through which it passed into a conductor of electricity, in which electricity can be studied in its simplest forms and under the simplest conditions. This perhaps did more than anything else to start physics off on the triumphal progress of the period to which we now come.

RADIOACTIVITY. The discovery of X-radiation was soon followed by that of other new radiations. The emission of X-radiation is a transitory affair which occurs only while an electric current is flowing, but these other radiations are permanent, being emitted continuously from certain substances. Foremost among these substances is uranium, a chemical element which was first isolated by Peligot in 1842 and, with an atomic weight of 238, was the heaviest element known until quite recently (p. 316). In February 1896, Professor Henri Becquerel found that a certain compound of uranium emitted a stream of radiation, continuously and of its own account, which resembled X-radiation in its deep penetration of matter, in affecting photographic plates, in exciting phosphorescence, and in turning gases through which it passed into conductors of electricity. It was soon found that this property resided in the uranium itself, and two years later Schmidt and Madame Curie found independently that thorium, another heavy element which Berzelius had discovered in 1828, possessed similar properties.

This inspired Mme Curie and her husband Professor Pierre Curie to undertake a systematic search for substances which showed 'radioactivity', as the new property was called. After a prolonged search which revealed nothing new, they tested the uranium-bearing mineral pitchblende from which their uranium had been extracted, and found that it was four times as radioactive as the pure uranium they had taken from it. Pitchblende must, then, contain some radioactive substance which was even more potent than uranium itself. Working with Bémont, they separated this in 1898 and called it radium. It proved to be another heavy element, of atomic weight 226, which was thousands of times more radioactive than uranium. They also found another element, polonium,* of atomic weight 210, which possessed similar powers, while in 1899 Debierne and Geisel discovered yet another, actinium, of atomic weight 227. All these radioactive elements were heavier than any hitherto known; broadly speaking, radioactivity is a property of the heaviest atoms, all elements heavier than lead (207) and bismuth (209) being radioactive.

The subject now engaged the attention of Ernest Rutherford, afterwards Lord Rutherford of Nelson, who had just been appointed Professor of Physics in McGill University.

* Alternatively described as radium F.

In 1899 he found that the radiation consisted of two distinct kinds of rays, which he called α-rays and β-rays. They were easily separated by their different penetrating powers, α-rays being reduced to half-strength after passing through $\frac{1}{50}$ mm. of aluminium foil, but β-rays only after $\frac{1}{2}$ mm. In 1900 Villard found that radium emitted a still more penetrating type of radiation, which he called γ-radiation. All radiations from radioactive substances were found to consist of one or more of these three types.

The next question was as to the nature and structure of these radiations. In 1899 Geisel, Becquerel, Curie and others examined the β-rays by the method which J. J. Thomson had used to measure the speed and the charge of the cathode-ray particles (p. 304), and found that they consisted of particles which were similar to these particles except that they moved much faster, some actually approaching the speed of light. Thus the β-rays were simply showers of high-speed electrons.

In the same way Rutherford found in 1903 that the α-rays consisted of positively charged particles moving at high speeds. These particles were only slightly deflected in either an electric or a magnetic field, which showed that their masses must be very great in comparison with their charges—the force dragging them round was small compared with the momentum carrying them on. He subsequently (1906) found that each particle had a mass which was more than 7000 times that of the electron, and a charge which was double that of the electron, but of opposite sign. The true nature of these particles was discovered three years later, when Rutherford and Royds shot a stream of them through a window of very thin glass, less than $\frac{1}{2500}$ inch in thickness, into a chamber from which they could not escape. They found that the gas helium (p. 249) was formed in the chamber, and continued to accumulate there so long as the α-particles continued to enter. Clearly the helium atom had α-particles as one of its constituents. The others were found to be the two electrons which were needed to neutralise the charge on the α-particle, and so make the whole atom electrically neutral. Thus the α-particle proved to be simply the helium atom robbed of its two electrons, or, as we should now say, the 'nucleus' of the helium atom (p. 315).

A more difficult problem was presented by the γ-rays, which refused to be deflected at all, either in an electric or in a magnetic field, and so might be either uncharged particles or electromagnetic waves. Ultimately, after much discussion, they were shown to be waves of very short wave-length—about a ten-thousand millionth part of an inch, or less than a hundred-thousandth part of the wave-length of visible light. Because of its shorter wave-length, X-radiation has a greater penetrating power than visible light, but γ-radiation, with its still shorter wave-length, has a greater penetrating power than either.

Meanwhile, a long series of new radioactive substances was being discovered, studied and isolated in rapid succession. In 1899, Rutherford noticed that a mass of thorium seemed to become less radioactive if a current of air was allowed to blow over it. The mystery was solved when he found that thorium emitted a heavy gas which was itself highly radioactive. This gas hung about the surface of the thorium so long as the surrounding air was still, but a breath of wind would sweep it all away, and then the radioactivity of the thorium seemed suddenly to diminish. Rutherford called this gas 'thorium emanation', and found that radium and actinium emitted similar emanations; that from radium proved to be a new element, and was called 'radon'. In 1900, William Crookes found that uranium yielded small quantities of a very radioactive substance which he called 'uranium-X', and two years later Rutherford and Soddy obtained a similar substance 'thorium-X' from thorium.

A vast number of other radioactive substances was now found, most of them showing intense but very transient radioactivity; clearly they had to pay for their great activities by short lives.

In 1902, Rutherford and Soddy made a general study of the decline of radioactive power, and found that it was governed by very simple laws. Samples of the same radioactive material would always lose the same fraction of their powers in a given time; it would take as long for the radioactivity of a sample to fall from 1000 to 900 as it subsequently took for it to fall from 100 to 90 or, later still, from 10 to 9. In mathematical language, the decay was exponential. But the rates of decay varied enormously from one substance to another, uranium losing half of its power in about 4500 million years, radium in about 1600 years, radon in about 3·8 days, and so on down to thorium C′, which fell to half-strength in perhaps a hundred millionth part of a second. No change of physical conditions could alter these rates of decay; the change seemed to come from inside the substance, and to be of the nature of atomic explosion or disintegration. Thus an atom had always the same chance of disintegrating, whatever its past history or present state might be. Here was a natural law of a kind hitherto unknown to science, and its implications were soon seen to be immense. From Democritus through Newton to the nineteenth century, science had proclaimed that the present was determined by the past; the new science of the twentieth century seemed to be saying something different—in the events now under consideration, the past had apparently no influence on the present nor the present on the future.

Rutherford and others now studied atomic disintegration in detail, and found that a radioactive substance went through a long sequence of changes, altering its chemical nature not once but time after time, passing through highly radioactive states very rapidly and less radioactive states more slowly, until it reached a final state in which it was completely and permanently stable. Uranium, for instance, went through no fewer than fourteen transformations before ending up as a new kind of lead which differed from ordinary lead in having the atomic weight 206 in place of the usual 207. These changes were accompanied by the ejection of α-, β- or γ-rays.

The emission of β- or γ-rays did not alter the mass of an atom appreciably, but when an α-particle was emitted, the atomic weight of course dropped by four. Thus uranium, which started with the atomic weight 238, passed through the atomic weights 234, 230, 226, 222, 218, 214 and 210, before ending as lead with atomic weight 206.

All these radioactive changes took place in the same direction—that of decreasing atomic weight; there was no traffic the other way. Thus radioactivity gave no encouragement to the view, as old as Democritus, that there were no changes in nature beyond a reshuffling of permanent atoms; it spoke rather of beginnings and endings, of a steady progress from a creation to a death, of an evolution which took place within a finite space of time. And it supplied a means of estimating this space of time.

THE AGE OF THE UNIVERSE. For many of the rocks of the earth's crust were found to contain small particles of radioactive material embedded in them, together with the various products of their disintegration. An analysis of the proportions of these various products revealed for how long the radioactive substances had been imprisoned in the rocks, so that an estimate could be formed of the time which had elapsed since the earth solidified.

The geologists had already formed their own estimates in a variety of ways, as, for instance, from the salinity of the oceans. The rivers are continually carrying water and salt down into the ocean; the water evaporates, but the salt does not, so that the ocean becomes continually more salty. From the present salinity of the ocean, the astronomer Halley (p. 195) had calculated that the earth must be several hundreds of millions of years old, and his estimate was confirmed by others, based on the rates of denudation of mountains and sedimentation of valleys—the rates at which mountains are worn away and valleys are silted up. But all these estimates depended on processes of which the speed was neither steady nor accurately known. Radioactivity now came forward with the magnificent gift of a clock of known and unvarying rate. The radioactive products in the oldest of terrestrial rocks were analysed, and were found to assign ages of nearly 2000 million years to these rocks, while a collection of meteorites analysed by Paneth and his co-workers revealed ages up to 7000 million years. Clearly the age of the universe was to be measured in thousands of millions of years.

This is the latest answer to a question which had been hotly debated for some time—the origin of the sun's energy. The ancients probably saw nothing surprising in the sun's continual outpouring of energy, but when Helmholtz came upon the principle of conservation of energy in 1857, he must have wondered where the sun found the energy for all its radiation. The only adequate source he could suggest was the sun's contraction. As a clock weight falls earthward, it provides energy for the continued motion of the clock; in the same way, thought Helmholtz, the shrinkage of the sun and fall of its outer layers centreward may provide energy for the continued emission of radiation. But Helmholtz calculated that this source could not provide energy for more than about 22 million years of radiation, whereas the geologists thought they had evidence that the sun had been radiating for hundreds of millions of years at least. Radioactive evidence not only supported the geological view very strongly, but also provided a clue as to a possible source of energy. For a sun made of pure uranium would radiate for a longer time than the sun's estimated age, and also more intensely. This suggested that the sun's energy might be of the same general character as radioactive energy, as now appears to be the case (p. 344).

THE STRUCTURE OF THE ATOM. These radioactive properties of matter not only compelled physicists to revise their conceptions of some very fundamental processes of nature, but also supplied them with a new working tool. The α-particle was in effect a high-speed projectile of atomic size and mass, and Rutherford saw that it could be used to explore the interior of the atom. In 1911 he suggested to two of his research workers, Geiger and Marsden, that they should pass a fusillade of α-particles through a layer of gas of such a thickness that a fair proportion of the projectiles were likely to hit an atom. They did this with sensational and entirely unexpected results. Most of the projectiles were found to pass clean through the gas without hitting anything at all, or even being deflected from their courses, thus dispelling at one blow the age-old picture of the atom as a hard solid chunk of matter; the atom was now seen to consist mainly of empty space. More unexpectedly still, a few of the particles were found to be deflected from their courses through very large angles. To quote Rutherford: 'It was almost as incredible as if you fired a 15-inch shell at a piece of tissue paper, and it came back to hit you.'*

The observed deflections conformed to a simple mathematical law, and from this it was possible to deduce the atomic structure which had caused them. Each atom, it was found, must contain a central core or 'nucleus' which was minute in size, but carried almost all the mass of the atom. It also carried a charge of positive electricity which caused the observed deflections of the bombarding α-particles, so that its amount could be calculated from the amount of the deflections. Moreover as the total charge on the atom is *nil*, this charge must just neutralise the charges on all the electrons of the atom, so that the number of these electrons could be calculated. For most substances it proved to be about half the atomic weight; an instance has already been provided by helium of atomic weight 4, which has two electrons in each atom. This led to a 'planetary' picture of the atom; the massive nucleus was like the sun, and round it the electrons revolved like planets.

Later, in 1913, H. G. J. Moseley and others found that the number of planetary electrons followed a very simple law:

* Rutherford, 'The theory of atomic structure', Essay in *The Background to Modern Science*, p. 69.

If the elements are arranged in order of increasing atomic weights, the number of electrons in the atoms are respectively 1, 2, 3, 4, 5, ... and so on—the series of integral numbers. Thus hydrogen, the lightest of all atoms, has only one electron in its atom; helium, the next lightest, has two; then comes lithium with three, beryllium with four, and so on. These integral numbers are called the 'atomic numbers' of the respective elements. At first there were a few gaps in the sequence of known elements, but these were rapidly filled by the discovery of new elements until a total of ninety-two elements was known, with atomic numbers ranging from 1 (hydrogen) to 92 (uranium). Quite recently, in 1940, the number has been increased by the further discovery of two new elements, neptunium and plutonium, of atomic numbers 93 and 94, to which two others of atomic numbers 95 and 96 were added, but not announced until 1946. When the simple law of the atomic numbers was first discovered, the problem of the structure of the atom seemed almost to be solved; few can have guessed how many hurdles still remained to be surmounted.

POSITIVE RAYS. The cathode particles are not the only mechanism for the transport of electricity through a gas; there is also a stream of positively charged carriers moving in the opposite direction. Goldstein discovered this in 1886 by the simple expedient of boring a hole in the cathode. Some of the particles which would otherwise have ended their journeys on the cathode, now passed through it, and were thus isolated for study. In 1898, W. Wien measured the charges and masses of these carriers by the method of electric and magnetic deflections (p. 304). Each particle was found to carry a positive charge equal in amount, but opposite in sign, to the charge on an electron, and to have the same mass as whatever atoms were in the discharge tube; indeed it was possible to measure the masses of these atoms by noting the deflections they experienced in known electrical and magnetic fields. Thus the carriers were simply atoms from which electrons had been removed—'positive ions' they were soon called.

It was now easy to see how electricity is carried through a gas. The electric forces which cause the current to flow pull the atomic positive and negative charges in opposite directions, until electrons are torn out of the atoms, leaving them as positive ions. The electrons and the positive ions now proceed to move in different directions under the influence of the electric forces—the negatively charged electrons from cathode to anode, and the positive ions in the opposite direction from anode to cathode. The two streams of particles form the cathode rays and the positive rays respectively.

ISOTOPES. The study of the positive rays had provided a new means of determining the masses of atoms, and hence the atomic weights of the elements, and this was soon found to be far more exact than the older method of the chemists. In 1910 Sir J. J. Thomson used it to measure the masses of the atoms of various simple substances. If the particles in a beam of positive rays were all precisely similar, and all moved at exactly the same speed, then all the particles would be similarly deflected by electric and magnetic forces, so that the beam would remain compact, and a photographic plate on which it fell would be marked at only a single point. Actually it is not easy to eliminate differences of speed, so that the beam spreads out and records a parabola on a photographic plate.* But when Thomson took the record of a beam of atoms of neon, he found two parabolas instead of one. The chemists had given the atomic weight of neon as 20·2; these two parabolas were found to represent atomic weights of 20·0 and 22·0, the former parabola being about nine times as strong as the latter. The startling inference was that neon did not consist of entirely similar atoms, but was a

* The magnetic deflection is inversely proportional to the speed of the moving atom, and the electric displacement to the square of the speed. Thus the electric deflection is proportional to the square of the magnetic deflection, and is at right angles to it, so that the curve is a parabola.

mixture of two different substances of atomic weights 20·0 and 22·0. To such groups of substances, Soddy gave the name 'isotopes' (ἰσο-τόπος), because they occupy the same places in the table of chemical elements—i.e. they have the same atomic number.

The subject was now taken up with great enthusiasm by F. W. Aston of Trinity College, Cambridge, who made a striking improvement in the apparatus and method, and then proceeded to analyse a great number of elements in a search for isotopes. He first resolved chlorine, of supposed atomic weight 35½, into a mixture of isotopes of atomic weights 35 and 37, the relative abundance being of course about 3 to 1 so as to result in an *average* atomic weight of 35½. After this, he and others produced results in a continuous torrent until nearly all the elements had been studied, their isotopes discovered, and their atomic weights measured with precision. The old idea had to be discarded that all the atoms of the same chemical element had been cast in the same mould. Each element has a definite atomic number which determines its chemical properties, but most elements consist of a mixture of atoms of different atomic weights.

It was found that if the atomic weight of oxygen is taken to be 16·00, then most of the atomic weights come out very near to integral numbers. Hydrogen has an atomic weight of 1·00837, with other isotopes of weights 2·0142 and 3·016; neon consists of three isotopes of weights 19·997, 21 (approximately) and 21·995; while krypton consists of a mixture of isotopes of weights 77·93, 79·93, 81·93, 82·93, 83·93 and 85·93. And most of the elements told a similar story.

We have seen how the hypothesis of Prout, that all atoms were aggregates of simple atoms of hydrogen, had fallen out of favour when it was found that the atomic weights were not all exact integers, and rightly so—no one could think of the chlorine atom as a conglomeration of 35½ hydrogen atoms. The new determinations of atomic weights went far towards removing this objection. For, although the new atomic

XII

Sir J. J. Thomson and Lord Rutherford. 1938

weights were still not exact integers, the small outstanding differences could easily be explained. We have seen (p. 295) how the principle of relativity requires every change of energy to be accompanied by a change of mass. If the distance between any two electric charges is changed, the energy of the combination is changed, and so also its mass. Thus the mass of any atom would change if its constituents were separated to a great distance apart, and if this were done for all atoms, their atomic weights might conceivably all become exact integers, so that Prout's hypothesis might again become tenable.

According to Rutherford, the hydrogen atom consisted of a single negatively charged electron and a positively charged nucleus, known as a 'proton', which carried a charge equal but opposite to that of the electron, the total charge on the atom thus being *nil*. For a time it was thought that every atom might consist only of protons and electrons—necessarily in equal numbers, since the total electrical charge on every atom is *nil*. If so, every atom would contain the ingredients of some integral number of hydrogen atoms, which is just what Prout had suggested. But this led to serious difficulties in connection with the magnetic properties of the nucleus, and other particles soon began to appear besides the proton and electron.

THE TRANSMUTATION OF THE ELEMENTS. We have seen how the alchemists devoted centuries of work to trying to transmute the elements, generally with the mercenary aim of changing base metals into gold. When their efforts met with no success, their aims were discredited, and even derided as impossible of achievement. The atoms, it was now thought, were permanent and unalterable structures; as they were now, so they had been made at the creation, and so they would stay until the end of time.

Then, in 1919, Rutherford performed an epoch-making experiment which showed that the projects of the alchemists had not been fantastic but were partly feasible. Not only so, but the prescription for changing the chemical nature of a substance was found to be amazingly simple: bombard the substance with α-particles. Rutherford first chose nitrogen for bombardment, and found that when an α-particle (or helium nucleus) of atomic weight 4 hit a nitrogen nucleus of atomic weight 14, the latter nucleus ejected a small particle which seemed likely to be a hydrogen nucleus. In April 1925, P. M. S. Blackett, then working in Rutherford's laboratory, arranged for the bombardment to take place in a Wilson condensation chamber (p. 304). In such a chamber, an electrified particle in motion can be made to leave behind it a condensation trail, rather similar to the condensation trail which an aeroplane leaves behind it in the upper layers of the atmosphere. This can be photographed, so that the path of the moving particle can be recorded. In many thousands of bombardments Blackett found that the nuclei simply rebounded from one another like so many billiard balls, but there were a few cases in which the combination of the nuclei of nitrogen and helium changed into a nucleus of oxygen of atomic weight 17 (one of the isotopes of oxygen) and a proton or hydrogen nucleus of atomic weight 1. The two nuclei of masses 14 and 4 had apparently exchanged three protons and an electron, and emerged as nuclei of masses 17 and 1. The process was in some ways suggestive of the radioactive process, but differed in being under control; instead of waiting for the trigger to go off by itself, the experimenter shook it by the impact of the massive α-particle, and the gun fired. There were the further differences that α-particles were absorbed instead of being emitted, and that protons were emitted, as they never were in radioactive transformations. This experiment opened up a vast field of research, which is still far from being exhausted, but there are few types of transmutation among the simpler elements which have not been studied in detail.

THE NEUTRON. In 1931, Bothe and Becker chose the light element beryllium for bombardment, and found that it emitted a highly penetrating radiation. As this could not be deflected by magnetic forces, it was at first thought to consist of γ-rays. The next year James Chadwick, then working in the Cavendish Laboratory, but now Professor at Liverpool, proved that it consisted of material particles of about the same mass as the hydrogen atom, but carrying no charge. He called these particles neutrons. They formed more effective projectiles than the α-particles, since, being uncharged, they were not repelled by the atomic nuclei.

It was soon conjectured that they might be normal constituents of atomic nuclei. A nucleus might contain protons equal in number to the atomic number of the element, thus giving the right charge to the nucleus, together with sufficient neutrons to bring the mass up to the atomic weight of the element. The addition or removal of neutrons would of course give isotopes. For instance, the nuclei of the three isotopes of hydrogen of atomic weights 1, 2, 3 would consist of a single proton together with 0, 1 and 2 neutrons respectively.

Experimental evidence soon confirmed this conjecture. Chadwick and Goldhaber broke the nucleus of the hydrogen atom of atomic weight 2 (called the 'deuteron') into a proton and a neutron, while Szilard split the nucleus of beryllium of atomic weight 9 into a nucleus of atomic weight 8 and a neutron.

These were only simple examples of a very general process, known as 'nuclear fission'—the splitting of a nucleus into smaller parts. Fermi and his colleagues in Rome bombarded uranium nuclei with neutrons, and thought they had obtained new radioactive elements heavier than uranium, until Hahn and Strassman showed in 1938 that they had merely broken the uranium nucleus into two smaller parts. Frisch and Meitner pointed out that a substantial part of the mass of the original uranium nucleus must have been transformed into energy, and Frisch confirmed this by showing that the parts of the shattered nucleus flew apart at explosive speeds.

A new chapter was opened in 1939 when nuclear fission was found to be accompanied by an emission of neutrons. This was important because if more neutrons were emitted than were absorbed, each newly emitted neutron might itself act as a bombarder, producing yet more neutrons, and so on indefinitely, thus producing an explosive of devastating power. This effect was found to be produced most simply by bombarding the nucleus of a rather rare isotope of uranium of atomic weight 235.

Here was the origin of that technique which has so far produced only the atomic bomb, but may conceivably be found in time to lead to industrial developments of the highest value. The transformation turns part of the mass of the nucleus directly into energy, and this may perhaps be utilised as a source for doing useful work, much as we now utilise the far smaller amounts liberated by the combustion of coal or the ignition of petrol vapour.

The energy stored in two electrically charged particles depends on their distance apart, varying inversely as the distance; if we reduce the distance between two particles to a millionth, we increase the energy we can obtain from them a million fold. Now in burning coal, or igniting petrol, or exploding nitro-glycerine, we are in effect rearranging charged particles which are at molecular distances apart, perhaps 10^{-7} or 10^{-8} centimetres. But in promoting nuclear fission we are rearranging charged particles which are at only nuclear distances apart, and these are of the order only of 10^{-13} or 10^{-14} centimetres. As these are only a millionth of molecular distances, the stores of available energy are a million times as great. Thus we must expect an atomic bomb to do about a million times as much damage as an equal weight of high explosive, and we may also hope for a corresponding increase of power if nuclear energy can ever replace chemical energy for peaceful purposes.

COSMIC RADIATION. So far our story of modern physics has been largely a story of the discovery of new radiations.

XIII

Launching of the Neher electroscope at
Fort Sam Houston, 1936

We must now describe yet another type of radiation which came to light in the early years of the present century. About 1902, a number of experimenters found that their electric instruments discharged themselves without any apparent reason, and conjectured that the cause must be some hitherto unknown type of radiation. This seemed to be present everywhere and it had a greater penetrating power than any type of radiation then known, for no thickness of metal could shield the instruments from its effects. At first it was thought to come out of the earth, but Gockel, Hess, Kolhörster, and later Millikan and his colleagues at Pasadena, found that their instruments discharged even more rapidly when sent up in balloons, while the opposite effect showed itself if they were taken down mines or sunk deep in radium-free water. It became clear that the radiation must come into the earth's atmosphere from outside.

The rays could not originate in the sun, for they were of the same intensity by night as by day; nor in our galaxy, for they still arrived in the Southern Hemisphere when the Milky Way was not visible. It seemed that the radiation must result from some general cosmic process, so that it became known as 'Cosmic Radiation'.

It is by far the most penetrating kind of radiation known, for it will pass through yards of lead. This makes it practically indestructible. For the average density of matter in space is so low (about 10^{-28} grams per cubic centimetre) that the radiation can travel for thousands of millions of years before encountering as much matter as there is in a sheet of lead one millimetre in thickness. As the universe is probably only a few thousands of millions of years old, this means that nearly all the cosmic radiation which has ever been generated is still travelling through space. Regener found that we receive as much of it as we receive of light and heat from all the stars except the sun; it breaks up about ten atoms in every cubic inch of air in every second. Averaging through the whole of space, it is probably the commonest kind of radiation in the whole universe.

For some years after the discovery of this radiation, there was much discussion as to its nature; did it consist of charged particles, or of electromagnetic disturbances, or of something different from either? It was pointless to test it in an ordinary laboratory magnetic field, for by the time it reached the laboratory it had passed through the whole of the earth's atmosphere, shattering every atom it met, and so becoming mixed up with a vast atomic debris of its own creation. The only magnetic field which could be used to provide a test was that of the earth itself, for the radiation passed through this before it got entangled in the atmosphere. This field would not deflect electromagnetic waves, so that if the radiation consisted of these, it would fall equally on all parts of the earth's surface. If, on the other hand, the radiation consisted of charged particles, these would be deflected in the earth's field, and so would fall unequally on the different parts of the earth's surface, the inequality in their incidence showing some reference to the earth's magnetic field.

From 1938 on, Millikan and his collaborators at the California Institute of Technology have been sending out expeditions to measure the strength of the radiation at different parts of the earth's surface, and they find that this is not uniform. Millikan and Neher interpret the observations as showing that at least 60% of the radiation must consist of charged particles, each moving with the energy that an electron acquires in falling through a voltage drop of from 2 to 15 thousand million volts. They make conjecture as to the origin of this energy.

We have seen how the mass of a particle increases as its speed of motion increases, so that an electron moving sufficiently fast might have the mass of a complete atom. This is not a mere fanciful conjecture, or even a deduction from theory, for Lauritsen and W. A. Fowler have found that a complete atom in the laboratory can transform itself into a pair of particles moving with such speed that their combined mass is equal to that of the original atom. Millikan has conjectured that cosmic radiation may contain particles of this type, their high speeds of motion endowing them with atomic masses. In 1943 Millikan, Neher and Pickering found that the observed radiation could be accounted for by the break-up of atoms of helium, nitrogen, oxygen and silicon—no mere random assortment of atoms, for it consists of some of the commonest atoms in space. If this finally proves to be the source of the cosmic radiation, it will provide an outstanding instance of the transformation of matter into radiation, although so far we can form no idea as to how or why this transformation takes place. We have already seen how a less complete transformation is used in the atomic bomb, and shall soon come upon another instance in the radiation from the sun and stars.

OTHER PARTICLES. The high penetrating power of the cosmic radiation implies a high shattering power, and actually the radiation shatters any atomic nucleus on which it falls. If the encounter occurs in a Wilson condensation chamber, the debris of the shattered nuclei can be examined by photographing the condensation trails of the various constituents (p. 304). In 1932, Karl Anderson, working at the Pasadena Institute of Technology, found that this debris contained particles of a hitherto unknown kind, which carried the positive charge of a proton, but the mass only of an electron. They were in fact positively charged electrons; Anderson called them positrons.

They have only a brief existence, coalescing with ordinary electrons almost as soon as they are born and disappearing in a flash of radiation, the energy of this flash being, of course, such that its mass is equal to the combined masses of the positron and the electron. Blackett and Occhialini of the Cavendish Laboratory suggested in 1933, and Anderson speedily verified, that the process is reversible; as the positron and the electron die in pairs, they are also born in pairs, matter thus being created out of energy.

So far these positive electrons were known only as products

of the cosmic radiation which came from the depths of space, but in 1934 Joliot found that similar particles were ejected from certain laboratory produced nuclei which were radioactive.

In 1937, it was found that the debris produced by cosmic radiation contained yet another new kind of particle—the 'mesotron' or 'meson'. This has the same charge as the electron, but a mass which has been variously estimated at from 40 to 500 times that of the electron, and so is intermediate between the masses of the electron and the proton. It is far from certain that this mass is always the same; there may be many kinds of meson.

All these various particles appear to emerge from atomic nuclei, which might thus seem to be mixtures of all these kinds of particles. But we do not know how far the particles have permanent and independent existences. Anderson has suggested that the neutron may not be a fundamental particle, but a proton and electron in combination—a sort of collapsed hydrogen atom; Rutherford and Aston found that its mass is about one-tenth of 1 % greater than the combined masses of a proton and electron. More generally, the proton and the neutron may be the same fundamental particle in different states, each transforming into the other by the emission of an electron or positron. This emission will of course produce a recoil, and to comply with the principles of the conservation of energy and momentum, other particles must be emitted at the same time; the purely hypothetical particles needed are known as the 'neutrino' and the 'anti-neutrino'.

In view of all this, it may be futile to try to assign any precise specification to the nucleus; there is not much point in discussing whether a cup of tea consists of milky tea sweetened with sugar or of sugary tea whitened with milk.

QUANTUM THEORY

The Kinetic Theory of Gases was mainly a creation of the nineteenth century. By the end of that century it had explained most of the properties of a gas by picturing it as a crowd of minute hard molecules darting about in space and frequently colliding with one another—to bounce off again on new paths, like billiard balls moving in three dimensions.

But the conception brought its difficulties. It was especially hard to understand why the darting and bouncing molecules should continue to dart and bounce—practically for ever. Billiard balls would not do so. They leave off darting over the table because they lose some of their energy of motion at every collision with another ball or cushion. This is transformed into heat energy, which is the energy of internal vibrations in the ball. Why do not the molecules of a gas transform their energies in a similar way?

One solution would be to assume that molecules could not vibrate, but this hardly seemed permissible. The spectrum of a gas was generally interpreted as evidence that its atoms or molecules were vibrating—just as the sound of a bell showed that the bell was vibrating. And later, when the atom was found to consist of a great number of electrical components, it seemed absurd to suppose that internal vibrations could not occur.

The difficulty was exhibited in sharp focus by a mathematical theorem known as the 'Theorem of Equipartition of Energy'. It showed that all possible motions of the molecules of a gas could be regarded as mouths waiting to be fed with energy, and competing for whatever energy was available. When two molecules collide, energy is transferred from one to the other, and the theorem showed that after a great number of collisions had taken place, the energy would be shared in a definite proportion. There would go to each molecule (on the average, taken over a small period of time), 3 units of energy for its motion through space, 2 units for each of its internal vibrations, and 0, 1 or 3 for its rotation, according to its shape and structure. If, then, there were many internal vibrations, most of the energy would go to feeding these. Actually experiment showed that most of it went to the bodily motion through space; in the simplest molecules of all—helium, neon, etc., in which the molecules are single atoms—it all went thus. Clearly something was wrong here, and as the theorem of equipartition was a direct logical deduction from the Newtonian system of mechanics, the fault seemed to lie with this.

The radiation from a red-hot body presented the same difficulty in a slightly different form. The theorem of equipartition showed that the radiation from such a body ought to consist almost entirely of waves of the shortest possible wave-length. Experiment showed a very different energy distribution in the spectrum.

The first move to end the deadlock was made by Max Planck, Professor in Berlin University, and subsequently in the Kaiser Wilhelm Institute. In an epoch-making paper which he published in 1900, he imagined all matter to consist of vibrators, each having its own particular frequency of vibration,* and emitting radiation of this frequency, just as a bell emits sound of its own frequency of vibration. This was completely in accordance with current ideas, but Planck now introduced the startling assumption that the vibrators did not emit energy in a continuous stream, but by a series of instantaneous gushes. Such an assumption was in flagrant opposition to Maxwell's electromagnetic laws and to the Newtonian mechanics; it dismissed continuity from nature, and introduced a discontinuity for which there was so far no evidence.

Each vibrator was supposed to have a certain unit of radiation associated with it, and could emit radiation only in complete units. It could never emit a fraction of a unit, so that radiation was assumed to be atomic. Such an assumption

* The frequency of a vibration is the number of vibrations which occur in a second.

naturally led to very different results from the Newtonian mechanics, but Planck was able to show that nature sided with him. His theory predicted the observed emission of radiation from a hot body exactly.

Planck described his units of radiation as 'quanta'. The amount of energy in any unit depended on the vibrator from which the unit came being equal to the frequency of its vibrations multiplied by a constant h, which is generally known as Planck's constant; this has proved to be one of the fundamental constants of the universe—like the charge on an electron or the mass of a proton. Through all the changes which the quantum theory has experienced—and they are many—h has stood firm as a rock, but we now associate it with radiation rather than with vibrators.

In 1905, Einstein attempted to represent the theory pictorially by depicting radiation as a flight of individual bullets of energy, which he called 'light-arrows'. These were supposed to travel in straight lines with the speed of light, each carrying just one quantum of energy, until they fell upon matter, when they were absorbed.

This, as Lorentz was quick to point out, was scattering the undulatory theory and all its triumphs to the winds, but the picture had much to commend it. When ultra-violet light, X-radiation or γ-rays pass through a gas, they break up some of its atoms, thus turning it into a conductor of electricity. It might be anticipated that the number of broken atoms would be proportional to the aggregate strength of the radiation which has passed through it. Actually the number is found to depend far more on the frequency of the radiation. Feeble radiation of high frequency may break up a large number of atoms, while intense radiation of low frequency may break up none at all—just as, with photographic activity, a little sunlight can fog our film, while a lot of red light (which is of low frequency) does no harm at all. This is readily explained if we picture radiation of high frequency, in accordance with Planck's ideas, as massive bullets, and radiation of low frequency as small shot. If a quantum has enough energy to break up the atom on which it falls, it does so, and the liberated electron carries off any residue of energy in the form of energy of motion. Actually the liberated electrons are found to move at precisely the speeds which this conception requires.

NIELS BOHR. In 1913 another great step forward was taken by Niels Bohr, who is now Director of the Institute for Theoretical Physics, Copenhagen. When the light from a gas is passed through a spectroscope and analysed, its spectrum is found to exhibit a series of lines each of which is associated with a quite definite frequency. Ritz had shown that these frequencies were the differences of others which were presumably more fundamental; if these latter were a, b, c, ..., then observed frequencies in the spectrum were $a-b$, $b-c$, $a-c$, ...

Bohr thought, however, that there must be something even more fundamental, namely, the quantities ha, hb, hc, ..., which are amounts of energy. His masterly central idea was that an atom could stand permanently only in states in which the energy had one or other of these values, but could suddenly drop from any such state to another of lower energy, giving out a quantum of energy in the process. If, for instance, the energy fell from ha to hb, the atom would emit radiation of energy $h(a-b)$, and this, in accordance with Planck's ideas, would constitute a quantum of frequency $a-b$, which was one of the frequencies in the observed spectrum.

Bohr next tried to illustrate these ideas by a study of the hydrogen atom, which he assumed with Rutherford to consist of a proton and a single electron revolving round it. He supposed the possible orbits of the electron to be those in which the angular momentum was an integral multiple of h, and found that the resulting values of ha, hb, ... led precisely to the observed spectrum of hydrogen. This spectrum, which had defied scientists for so long, seemed to have yielded its secret immediately to the new concepts of the quantum theory. Bohr next extended his investigation to the spectrum of ionised helium with entirely satisfactory results, but his theory was found to fail for the spectra of atoms more complex than helium.

The new ideas were open to the even more serious objection, in the eyes of many, that they were inconsistent with the undulatory theory of light. Mathematicians had to try to weld the old undulatory theory into a single whole with the new theory which seemed quite incompatible with it.

HEISENBERG, BORN AND JORDAN. Little progress was made until 1925, when a great forward step was taken by Werner Heisenberg, who had worked with Bohr at Copenhagen and was subsequently Professor of Physics in the University of Leipzig, and by Max Born who, after being Professor in Berlin, Frankfurt and Göttingen, became Professor of Natural Philosophy in the University of Edinburgh. Heisenberg thought that the imperfections of Bohr's theory must result from imperfections in his picture of the atom. When an atom was broken up, electrons came out of it, but these ingredients might have changed their attributes in the process of breaking up; the electron tied inside the atom might be something quite different from the electron free in space. Heisenberg accordingly discarded all unproved conjectures as to the existence of particles, quanta of energy, waves of light, etc., which could not be observed, and concentrated on the 'observables', the existence of which could not be doubted; these consisted of nothing but spectral lines, of which the frequency and intensity could be measured. Following Heisenberg along these lines, Born and Jordan devised a system of laws which proved to be in exact agreement with observations on atomic spectra. The new system of laws is generally known as the 'Matrix Mechanics'.

Ordinary algebra deals with ordinary simple quantities, which it denotes by simple symbols such as x, y, z. If a group of quantities is intimately connected, it may sometimes be convenient to deal with the whole group *en bloc*, denoting it by a single letter. A group of one special kind, which need not be specified here, is described as a matrix. A number of mathematicians had studied the properties of matrices long before they were known to be of importance to atomic physics, and had formulated rules for their manipulation. For instance, if p denoted the group $a_1, b_1, c_1, \ldots$, and q the group $a_2, b_2, c_2, \ldots$, it is obvious to take $p+q$ to denote the group $a_1+a_2, b_1+b_2, c_1+c_2, \ldots$. There are analogous conventions as to the meaning of $p-q$, pq, p^2, $1/p$, and so on.

A great advance was made when Heisenberg, Born and Jordan and others showed that the workings of nature on the atomic scale conformed to laws which were Newtonian in form, except that the simple algebraic quantities of Newton had to be replaced by matrices. If the generalised co-ordinates and momenta which appear in the canonical equations (p. 235) are replaced by suitably chosen matrices, the laws so obtained appear to govern the whole of atomic physics. Bohr's plan had been to retain the particle-electron, and to modify the Newtonian mechanics; Born and Jordan retained the Newtonian mechanics (at least in form) and modified the particle electron, replacing it by something which was unknown, but was certainly more complex than a simple particle. This unknown something they could only specify mathematically. The Newtonian mechanics were valid up to the boundaries of the atom, after which the quantum mechanics took over. In the free space outside the atom, and also in the outer confines of the atom, the new electron reduced to a simple particle, and the new scheme of Heisenberg, Born and Jordan became identical with the older scheme of Bohr, both reducing to the Newtonian mechanics.

Matter was now seen to be something more complicated than a collection of particles. Bohr's theory had been a final attempt to interpret it as particles, but it had become clear that something more refined was needed to explain the internal workings of the atom. The necessary new conceptions did not permit of representation in mechanical terms; indeed, they could not be represented in space or time at all. The conjecture of Democritus that the universe could be explained as a void inhabited by particles had served science well for 2400 years, but the time had now come to discard it, and with the failure of Bohr's theory, the conception of the universe as a structure of particles existing in space and time had to be dismissed from science.

DE BROGLIE. While all this was in progress, other attempts were being made, on quite other lines, to discover the true scheme of natural laws. In particular, Louis de Broglie of Paris had in 1924, given a new turn to the discussion guided by an optical analogy. Reflecting on how the primitive theory of optics, which had regarded light as rays travelling through space in straight lines, had been compelled to give place to the more refined wave-theory, he had thought that the theory of the moving electron might be improved in a similar way. He began to think of the moving electron as a train of waves, and showed how the principles of the quantum theory made it possible to assign frequencies and wave-lengths to the waves. In 1927, Davisson and Germer, almost by accident, let a shower of rapidly moving electrons fall on the surface of a crystal, and found that the electrons were diffracted and formed the same kind of pattern as X-radiation formed under similar conditions. As X-radiation was believed to consist of waves, this suggested that there was something of an undulatory nature about the electron. The next year, G. P. Thomson, the son of Sir J. J. Thomson, and then Professor of Natural Philosophy in the University of Aberdeen, passed a shower of electrons through a very thin film of metal, and observed a similar effect, the wave-length and frequency of the waves being exactly those required by the quantum theory. It began to look as though matter consisted of waves rather than of particles.

SCHRÖDINGER. In 1926, Erwin Schrödinger, then Professor in the University of Berlin, applied the same ideas to the motion of the electron inside the atom, substituting a train of waves for each of the electrons postulated by Bohr's theory. This theory had permitted an electron to move only in certain orbits in the atom, and Schrödinger now showed that the permitted orbits were those which just contained an integral number of complete waves, so that the wave pattern joined up neatly to complete the circle. In this way he arrived at a mathematical specification which seemed to make a fundamental explanation of spectra possible. So far the waves had only been mathematical abstractions; their physical interpretation was now explained by the 'principle of uncertainty' or indeterminacy which Heisenberg now introduced.

THE PRINCIPLE OF UNCERTAINTY. It is a commonplace in science that the best results cannot be obtained from poor instruments; the more precise our instruments, the more exact our results can be. If we had instruments of perfect precision at our disposal, we should be able in principle to draw a perfectly exact picture of the material universe; we could say, for instance, 'here at this exact point of space and at this precise instant of time, there is an electron, moving at such a speed'. But our instruments are themselves part of the universe which we wish to explore, and they share its defects, one of which, for exploratory purposes, is its atomicity. Because matter and radiation are both atomic, we can never obtain perfectly precise instruments for our investigations, only clumsy blunt probes which do not permit of our drawing clear-cut pictures of anything. The smallest mass we can command is that of the electron; the smallest amount of energy we can liberate is that of a complete quantum. The impact of either an electron or a quantum upsets the part of the universe we are trying to study, and replaces it by something new, which in turn evades precise study in a similar way. Such are the ideas underlying the 'Principle of Indeterminacy' which Heisenberg introduced in 1927.

Heisenberg showed that this coarse-grainedness of nature makes it impossible in principle to fix both the position and the speed of an electron with perfect precision; if we reduce the uncertainty in one, it automatically increases in the other, and the product of the two uncertainties can never be reduced to below a certain minimum value. This minimum value is a simple multiple of Planck's constant h, as is indeed natural. For as this constant specifies the atomicity of radiation, we must expect it also to specify the indefiniteness in our knowledge which results from this atomicity. Thus our measurements of position and speed must be regarded as indicating probabilities rather than certain facts.

In 1926 Born showed that the mathematical waves of de Broglie and Schrödinger can be interpreted as graphs exhibiting the probabilities of electrons being at the various points of space; no waves at a point means zero probability, weak waves means slight probability, and so on. The waves do not exist in ordinary three-dimensional space, but in a purely imaginary multi-dimensional space, and this alone shows that they are mere mathematical constructs, which can have no material existence. Nevertheless, their propagation according to definite known equations gives a perfect account of the happenings inside the atom, or at least of the radiation which comes out of the atom. The waves may equally well be interpreted as diagrammatic representations of our knowledge of the electrons concerned, for which reason they are sometimes described as 'waves of knowledge'.

A full discussion of these results made it clear that matter could not be interpreted either as waves or as particles, or even as waves plus particles. Matter shows some properties which are inconsistent with its being waves, and some which are inconsistent with its being particles. It became generally agreed that it must be interpreted as something which in some of its aspects reminds us of particles, and in others reminds us of waves, but for which no intelligible model or picture can be constructed. The waves have to be waves of probability, or waves of knowledge—the two interpretations are equivalent—while the particles are wholly material, being indeed the standards from which we get our ideas of materiality. But the universe perceived by our senses consists of matter and radiation. If matter was as we have just said, what of radiation?

As we have already seen, there were at this time two apparently inconsistent views of radiation. One regarded it as waves—the electromagnetic waves of Maxwell; the other regarded it as particles—the 'light-arrows' of Einstein, which we now call 'photons'. Clearly there was a dualism here just like that found in interpreting matter. And the explanation proved to be the same: we can never know where a photon is precisely; it is all a matter of probabilities, and just as the waves of de Broglie and Schrödinger exhibit the probabilities of electrons being in various places, so the waves of Maxwell's electromagnetic theory and of the undulatory theory of light—the waves which we usually describe as light-waves—may be interpreted as waves expressing the corresponding probability for photons.

DIRAC. In 1930, P. A. M. Dirac, now Professor of Mathematics in the University of Cambridge, published an important book, *Quantum Mechanics*, which aimed at putting the whole theory in a consistent mathematical form, and at unifying the various theories then in the field. He produced a mathematical theory of a very abstract kind which, as he showed, contained the matrix-mechanics and the wave-mechanics as special cases. Its basic conception was that the fundamental processes of nature cannot be depicted as happenings in space and time; beyond anything that we can observe, there is a substratum of events that do not permit of such representation. The observation of these events is a sort of process which the events may undergo, and by which their form is changed; it brings them, so to speak, to the surface of the substratum, where they permit of representation in space and time and may now affect our instruments and senses.

This brings us to the frontier of present-day knowledge of quantum theory, and here for the moment progress seems to be checked. There are still many unsolved difficulties, some of which may prove to be fundamental, and innumerable points of detail on which more knowledge is needed. Some physicists think that the present difficulties may soon be surmounted by minor modifications of existing theories. Others, less optimistic, think that something completely new and fundamental still remains to be discovered, or that some great simplifying synthesis is still to come, which will make all things clear. Every new explanation of nature has seemed bizarre or irrational at first; our tendency is to expect a mechanical universe conformable to our experience of our everyday man-sized world, and the further we go from this man-sized world, the stranger the world in which we find ourselves appears. To this quantum theory forms no exception.

OTHER DEVELOPMENTS IN PHYSICS

We have now mentioned most of the major new developments in the physics of the present century, but it would create a very wrong impression if we were to end our discussion here. For an enormous amount of work has been devoted in recent years to developing and extending the work of the nineteenth century, particularly in the direction of widening the limits of the temperatures and pressures available in the laboratory, and studying the properties of matter under the new conditions thus made accessible.

Perhaps the most striking pushing back of boundaries has been in the direction of lower temperatures. The two difficulties are of course, first to get the substance under study down to the required low temperature, and second to keep it there.

Early in the nineteenth century, Gay-Lussac thought he had shown that when a gas expanded freely and did no external work, it experienced no change of temperature, and therefore no change of energy. Later Joule, Lord Kelvin and others found that this was not completely true; a gas which passed through a porous plug into a vacuum beyond cooled slightly, because work was consumed in overcoming the mutual attractions of the molecules of the gas. At low temperatures, the fall of temperature may be substantial, and by passing a gas through a porous plug time after time, the temperature can be continually lowered. By use of this principle, James Dewar succeeded in liquefying hydrogen in 1898, at a temperature of about $-252{\cdot}7^\circ$ C. After this, many other gases were liquefied in rapid succession until finally, when Kamerlingh Onnes liquefied helium in 1908 at a temperature of $4{\cdot}22^\circ$ above the absolute zero, the last gas had become available in the liquid state.

The second difficulty was overcome by the 'vacuum flask' of James Dewar. A layer of air, as we all know, is an exceedingly good non-conductor of heat; we wear clothes to keep such a layer of air next to our skins. The less dense the air the better the heat insulation, so that a perfect vacuum is the ideal non-conductor. The Dewar vacuum flask utilises this principle; it consists of a flask with two walls which are separated by a vacuum; it is the ordinary 'thermos flask' of commerce, which we use to keep our soup or coffee warm; Dewar used it to keep his liquid air or other substances cold, and it has proved of the utmost service in the study of matter at low temperatures.

Kamerlingh Onnes did not stop when he had liquefied helium, but with a band of co-workers at Leyden, he proceeded to ever lower temperatures, soon approaching to within a degree of the absolute zero. Magnetic means are now available for passing to within less than a two-hundredth of a degree from the absolute zero.

Access to these low temperatures has opened up a vast new field—the study of matter which is almost undisturbed by the heat-motion of its molecules. Here many of the properties of matter prove to be completely different from those we know in ordinary life. Kapitza has shown that the magnetic properties of most substances change in a remarkable way at low temperatures. In 1935, Keesom and Rollin found that, at temperatures within 2·19° of absolute zero, helium changes its character so completely as to become almost a different gas, which is now described as helium II. This gas has almost no viscosity and is an amazingly good conductor of heat. But perhaps the most surprising property of all is that known as 'superconductivity'. When a metal which exhibits this property is cooled below a certain critical temperature which is characteristic of the metal, it becomes an almost perfect conductor of electricity. At any ordinary temperature, a current which is induced in a circuit and is not maintained, will disappear in a fraction of a second; but at a sufficiently low temperature, the current may continue of its own momentum for hours or even days. This last effect has been known for years, but there is still no satisfactory theory to account for it, and the same may be said of most of the low-temperature phenomena; they open up problems rather than add to the main body of science, and as such need only a brief mention in a history of science.

In the direction of high temperatures there is less to be recorded; temperatures approaching to 20,000° C. have been attained, but it has not so far proved possible to do much with them. Nature provides much higher temperatures for us in the stars, and performs experiments with them in her own way, and it is perhaps through astrophysics that we shall best gain a knowledge of the properties of matter at the highest temperatures.

Much the same must be said of extremes of pressure. It is now possible to attain any pressure within the range from a thousand millionth of an atmosphere to about 100,000 atmospheres, but again far wider limits are found in astrophysics, and it may be that the effects of extreme pressures are best studied here.

ASTRONOMY

Astrophysics

Astrophysics is the study of the physical constitution of the stars. It was conceived as an offshoot from spectroscopy when John Herschel suggested in 1823 that a study of a star's spectrum might disclose the chemical composition of the star, and for nearly a century this study seemed to be the central task for the new science.

In 1867, Father Secchi of the Vatican Observatory in Rome divided stellar spectra into four main classes. It is now usual to employ a more detailed classification which was devised by Harvard Observatory. It is found that the various observed types of spectra can be arranged in a continuous sequence, each type merging gradually into its neighbour. The reason is that a star's spectrum depends almost entirely on the temperature of its surface layers, and the continuous series is simply one of decreasing temperature. Stars of bluest colour come first in this series, and stars of red colour last, for the hotter a star is, the more blue its colour.

The various types of spectra show the characteristic lines of different chemical elements, one showing helium in great strength, the next showing hydrogen, and so on. It was at first imagined, somewhat naïvely, that the stars consisted mainly of those elements which showed most strongly in their spectra. When the spectral types were found to form a linear sequence, it was thought that this might represent the different stages in the star's evolution, and pictures were drawn of stars consisting mainly of hydrogen at the beginnings of their existences, and gradually changing into heavier elements as they aged. All this disappeared when Saha gave the correct interpretation of stellar spectra in 1920 and 1921: At different temperatures, the different chemical elements of a star show up in strength. Thus at a temperature of say 10,000°, the atoms of some substances, but not all, will be actively engaged in emitting radiation; if the temperature falls to half, these substances will leave off emitting radiation, while others will take their place. On the old interpretation of stellar spectra, a star which was suddenly cooled from 10,000 to 5000° would have seemed suddenly to change the hydrogen, helium and iron of which it was made into calcium, carbon, etc. We know now that all stars are formed out of much the same mixture of elements, but show different spectra because their surface layers are at different temperatures. In this way the spectrum is useful as an indication of temperature. But this is far from being its only use; it also yields information as to the motion of a star in space, the intrinsic brightness of a star, its mass and the composition of its atmosphere.

GIANT AND DWARF STARS. As most stars are so distant that no telescope can make them appear as more than mere points of light, it is impossible to determine their diameters by direct measurement. But a star's spectrum tells us its temperature, so that we know how much radiation it emits per square metre of its surface; we also know its total emission of radiation from the amount of light we receive from it. A simple division now tells us its total area of surface, and hence its diameter.

In 1913, Professor Hertzsprung of Leyden calculated some stellar diameters by this method, and found that the coolest stars of all, those of deep red colour, fell into two distinct classes, stars of very great brightness and stars of very small brightness; he called them giants and dwarfs. Stars of intermediate size were non-existent. Shortly afterwards, Professor H. N. Russell found that the same was true, although to a lesser degree, of all the cooler types of stars. Passing to hotter stars, the gap between large and small stars persisted over a considerable range, and finally disappeared—the two branches of large and small stars merged into one.

Russell represented all this in a diagram—the famous 'Russell diagram'. He put the reddest stars of all at the extreme right-hand of the diagram and the bluest at the extreme left, stars of other colours being put in the appropriate positions. He also put the stars at different heights in the diagram to represent their different brightnesses and so of course their different sizes, the brightest of all being put at the top, the faintest at the bottom, with intermediate stars again in their appropriate intermediate positions. When he did this Russell found that his diagram assumed the form (roughly) of a V lying on its side, thus: <. The upper branch consisted of course of giant stars, the lower of dwarfs; Hertzsprung's gap was that between the two arms at the extreme right.

Russell suggested an evolutionary interpretation of this. In brief, he thought that as a star aged it slipped downwards from the top to the bottom of the <, thus changing first from red to blue, and then returning back from blue to red, while its intrinsic brightness steadily decreased. A star had generally been supposed to start its life as an immense and comparatively cool mass of nebulous gas. Now the American J. Homer Lane had shown in 1870 that, as such a mass of gas lost energy by radiation, it would contract, but would at the same time get hotter. Russell supposed that as a result of this contraction and heating, a mass of nebulous gas would first become a giant red star of immense size, and then would pass through the sequence of the giant stars, becoming continually hotter and smaller until finally its density became comparable with that of water. The mass was now so far from the gaseous state that Lane's law no longer applied. Russell identified this stage with the junction of the giant and dwarf branches—the tip of the <—and imagined that from now on the star both shrank and cooled, passing through the sequence of dwarf stars until finally it vanished into darkness. For a time this train of ideas seemed to provide a satisfactory account of the observed varieties of stellar spectra and diameters, as well as giving a plausible account of the evolution of the stars. But it became untenable in 1924, when Arthur Eddington, Plumian Professor in the University of Cambridge, showed that the luminosity of an ordinary star depended mainly on its mass. Thus so long as the mass of a star remained approximately the same, the star could not become appreciably brighter or fainter. And there is no reason for expecting any substantial change in the mass of a star during the few thousands of millions of years which form the life of the average star, so that the luminosity of an ordinary star must remain appreciably constant. In the light of this new knowledge, Russell's scheme of evolution became untenable.

STELLAR INTERIORS. So far observation and theory had been concerned only with the surfaces of stars; interest now passed to their interior mechanisms, which only theory could reach. In 1894, R. A. Sampson of Edinburgh had shown that heat would be transferred inside a star by radiation rather than by conduction, but his study of stellar interiors was invalidated through his assuming a wrong law of radiation. In 1906, Professor Schwarzschild of Göttingen attacked the same problem, and gave correct formulae for the transfer of energy. From 1917 on, Eddington made an intensive study of stellar interiors, and was able to provide a theoretical explanation of his law connecting a star's mass with its luminosity. He obtained one result which proved to be of great importance: The temperatures at the centres of the ordinary dwarf stars are all nearly equal—about 20,000,000° —and depend only slightly on the size and mass of the star.

This suggested a new scheme of stellar evolution. A star could be supposed to begin as a mass of cool nebulous gas, and to contract until its central temperature reached the 20,000,000° just mentioned. It would now have become an ordinary dwarf star, and would retain its present luminosity and size for a great time. Indeed, the preliminary stage of contraction would have occupied only a few million years (p. 314), so that thousands of millions of years would remain, during which the star ought to radiate energy without much change of size or condition. This suggested that energy must somehow be liberated in the star's interior as soon as the temperature reached 20,000,000°. The question was: how?

We have already mentioned the unproved conjecture that the evolution of a star might be accompanied by a transmutation of its substance from light elements to heavy. The experiments of Rutherford and others had shown that there was nothing intrinsically impossible in such an idea. As the atomic weights of most elements are not exact integers, every transmutation is likely to be accompanied by a small loss or gain of mass; if mass is lost, the corresponding energy will be set free in the form of radiation. Perrin and Eddington had both suggested that the transmutation of hydrogen into heavier elements would provide about the right amount of radiation for the thousands of millions of years which now appear to be the normal lifetime of a star.

THE SOURCE OF STELLAR ENERGY. All this seemed to show that a star might obtain the energy for its radiation from a transmutation of its substance, but it took some time to discover the actual process. In 1938 and 1939 Bethe, Gamow and Teller and others proposed a scheme, based on preliminary calculations by R. d'E. Atkinson and Houtermans (1929) which now is thought to provide the most probable description of what happens.

In brief, the scheme supposes four protons to combine to form a helium nucleus with the assistance of a carbon nucleus which expedites the process by its presence, but ultimately emerges unchanged—as the chemists say, it acts catalytically. The carbon nucleus (of atomic weight 12) first captures a proton and combines with it to form a nucleus of atomic weight 13, this being one of the isotopes of nitrogen. This decays into a carbon isotope of weight 13 by emitting a positron. Two more protons are captured in succession and another positron emitted, forming a nitrogen nucleus of weight 15. Finally this reacts with another proton, but instead of producing an oxygen nucleus of weight 16 it forms two nuclei of weights 12 and 4. The former is the original carbon nucleus, which is now restored uninjured to the star; the latter is a helium nucleus which has a smaller mass than the four protons which have been hammered together to form it. The difference of mass, which is approximately that of 0·028 proton, has been set free as radiation. Although the scheme is complicated in the sense that many transmutations must occur before the final product is reached, yet it is confirmed by laboratory observation at practically every stage. Sen and Burman have calculated (1945) that it gives the right emission of radiation for the sun if this is supposed to consist of 35 % of hydrogen by weight, which agrees well enough with previous estimates, and if matter at the centre of the sun is 45 times as dense as water and has a temperature of 20·2 millions of degrees.

It is now generally supposed that the ordinary dwarf stars have been formed in some such way as this. But before reaching its present state, the matter at a star's centre must have passed through all temperatures up to about 20,000,000°. And there are other nuclear reactions—those of protons and deuterons (p. 321) with one another, and with the nuclei of the light elements lithium, beryllium and boron—which are known to take place at temperatures well below 20,000,000°. As the matter of the dwarf stars must have passed through the temperatures at which these transmutations operate, it would in any case seem likely that the stars would pause for some time in their evolution at these lower temperatures, their radiation being provided by the using up of the nuclei of these light elements. Actually, as Gamow and Teller have pointed out, there are distinct groups of stars which have just about the central temperatures necessary for these reactions to occur. The group of stars known as red giants (p. 341) have the right central temperature for the reaction of deuterons with other deuterons and with protons. Another group, that of the stars known as Cepheid variables (p. 349), is less clearly defined, but Gamow and Teller think this is formed by the overlapping of three separate groups, their central temperatures being those needed for the reactions of protons with the nuclei respectively of lithium, beryllium and boron of atomic weight 11. Finally comes a group of variable stars known as cluster variables, with central temperatures appropriate to the reaction of protons with boron of atomic weight 10.

Thus it seems permissible to picture a star as passing through the sequence of states in which deuterons and the nuclei of the light elements lithium, beryllium and boron are all transmuted in turn until the supply of these elements is exhausted. After this, the star will contract until its centre reaches the temperature at which the protons interact with carbon nuclei. In the earlier interactions the light elements had been used up, while some at least of the hydrogen survived. Now the hydrogen is itself used up, but the carbon survives, and in time a stage must be reached in which there is no longer enough hydrogen to maintain the radiation of the star.

When this stage is reached, the star must again contract, and its central temperature will begin by rising. It may be that other reactions will provide further energy for radiation, but in any case the star seems likely to end as what is known as a 'white dwarf'. In these stars, the matter is packed very closely, so that their density is many thousand times that of water; their diameters are small—sometimes no larger than the earth—and although they emit but little radiation, the temperatures of their surfaces are high, rising to about 70,000° for the central stars of the planetary nebulae. These stars emit so little radiation that even the energy set free by their contraction under gravitation is sufficient to provide for their radiation through a long life. In the end they must gradually use up all their stores of energy, and fade away into darkness.

Observational Astronomy

Throughout the nineteenth century observational astronomy, like physics, had made continuous progress along well-established routes; with the turn of the century, again as in physics, new methods and new ideas were introduced, which expedited progress enormously.

First and foremost, the twentieth century is the era of giant telescopes—or so at least it appears to us. It has been said that the history of astronomy is one of receding horizons and, in the opening decades of our century, thanks largely to the rapidly increasing size and power of telescopes, the horizon receded at an amazing pace. When the nineteenth century ended, astronomical knowledge was almost entirely confined to the solar system. The observatories gave their attention mainly to the motions of the sun, moon and planets, and to the appearances of the last, while outside the observatories popular lecturers tried to make our blood run cold by telling us how long express trains would take to pass from one point of the solar system to another. Little was known about the stars except the distances of a few of the nearest, and even these distances were very inaccurate.

Gradually the centre of interest shifted from planets to stars and later on, at least for many astronomers, from stars to nebulae. Each of these steps represented a million-fold deeper penetration into space, for the nearest stars are almost exactly a million times as distant as the nearest planets, while the nearest of the nebulae are at nearly a million times the distances of the nearest stars.

STELLAR DISTANCES. Added to the mere increase in the light-gathering power of telescopes which resulted from increase of size, were technical improvements which perhaps played an even more important part. One of the most valuable of these was the application of photography to astronomy. It showed its value in the determination of stellar distances, first by Schlesinger at Allegheny in 1902, then by A. R. Hinks and H. N. Russell at Cambridge in 1905, and soon afterwards in most of the observatories of the world. Before the introduction of photography, the distance of a star was found by first measuring out the angular distances separating it from a number of fainter, and so presumably very distant, stars, and then noticing how these distances changed as the earth moved in its orbit round the sun. If a star was not itself in motion relative to the sun, the angle through which it had moved across the background of faint stars would give a measure of its distance—of course in terms of the diameter of the earth's orbit. Most stars are in motion relative to the sun, but this relative motion is easily detected and allowed for, in a way to be explained below. Photography now replaced the tedious and uncertain measurement of angles in the sky by simple measurements of distances on a photographic plate; it was only necessary to take photographs of the same small part of the sky at suitable intervals of time, and the thing was done. The distances of thousands of stars have been measured by this method, but it is useless for very distant stars because their apparent motion in the sky is too small to measure; other methods must be found for these. Still the accurate distances of a few hundred stars provide a sort of yardstick in terms of which greater distances can be measured by other means, and we shall soon see how the astronomer now has at his disposal a selection of yardsticks of different kinds and lengths, with which to measure distances ranging from $4\frac{1}{4}$ light-years, the distance of the nearest stars, up to about 500 million light-years, which is the distance of the furthest objects we can see in space.

We have seen how Newton tried to estimate stellar distances by assuming that the stars were 'standard beacons' each being of the same intrinsic brightness (or 'luminosity') as the sun. Such an assumption is not permissible to-day, for we know that the luminosities of the stars range from less than one 300,000th of that of the sun up to 300,000 times that of the sun. But certain definite classes of objects may still be used as 'standard beacons', their luminosities all being equal and of known amount, so that the distance of such an object may at once be deduced from the faintness of its appearance.

Foremost among such objects are certain types of variable stars—stars which do not shine with a steady light, but fluctuate in brightness. The most interesting and useful of all are the stars known as Cepheid variables, being named after their prototype, the star δ Cephei. These fluctuate perfectly regularly with definite periods, and in a very distinctive way —a rapid brightening of the light being followed by a much slower decline—so that they are easily recognised.

In 1912, Miss Leavitt of Harvard made a study of the Cepheid variables in the Lesser Magellanic Cloud, a vast group of stars out beyond the confines of the Galactic System, and found that all which had the same period of light fluctuation appeared equally bright. The apparent brightness of a star generally depends both on its distance and on its intrinsic luminosity, but in this case all the stars of the cloud were approximately the same distance. Clearly Miss Leavitt's result could only mean that all stars of the same period were also of the same intrinsic brightness; in other words, Cepheids of any single assigned period could be used as 'standard beacons'. As the distances of a number of Cepheids have now been measured in the way already explained, it is possible to deduce the intrinsic brightness of Cepheids of every period, and hence to deduce the distance of any object in which Cepheids can be recognised. These stars have provided one of the most useful of yardsticks for the survey of the depths of space.

STELLAR MOTIONS. Photography soon proved itself equally useful for the study of the motions of the stars in space. If two plates are taken at exact yearly intervals, the disturbing effect of the earth's motion round the sun is eliminated, and any motions of the stars across the face of the sky must represent real motions relative to the sun. The best results are naturally obtained by measuring the motion over a long interval of years, and observations which Bradley made of stellar positions in 1755 have proved specially useful.

This method will, of course, only inform us as to the motion of a star in directions round the sun; it cannot disclose motion directly towards, or directly away from, the sun. As far back as 1868 Sir W. Huggins had suggested that it might be possible to determine such motions from a study of the star's spectrum; any motion along the line of sight would produce a displacement of the lines in the spectrum, and the speed of motion of the star could be deduced from the amount of this displacement. In the early years of the century, W. W. Campbell, Director of the Lick Observatory, used this method very successfully, and the radial velocities of a great number of stars were soon known with good accuracy.

At this time it was generally imagined that the stars were engaged in purely random motions, and the main interest of the problem of stellar motions was to determine the motion of the sun through space. But when Kapteyn of Groningen examined stellar motions statistically, he found that they were not pure random motions, but showed signs of conforming, although only imperfectly, to a definite plan. In 1905 he announced that the stars in the proximity of the sun could be divided into two distinct streams which moved through one another in opposite directions. A few years later Schwarzschild, Eddington and others showed mathematically that the observed motions could be interpreted in other ways than as those of two streams of stars. Of late years Oort, Lindblad, Plaskett and others have made further statistical studies of the motions, and have discovered a new and remarkable regularity. It has often been remarked that only the motion of the stars could save them all from falling together, under their mutual gravitation, at the centre of the system; in 1913 Henri Poincaré had calculated that a rotation with a period of about 500 million years would suffice to preserve them from such a fate. In the same year Charlier announced that the plane in which the planets described their orbits round the sun seemed to be slowly turning about in space, or at least against the distant background of the Milky Way, with a period of about 370 million years. Now dynamical theory requires that this plane should continually preserve the same direction in space; for this reason it is known as the 'invariable plane'. Eddington at once suggested that it was not this plane, but the star-system of the Milky Way, which was turning. It is now clear that this rotation is a reality, although it is not a simple rotation like that of a cartwheel. It rather resembles the motion of the planets round the sun, where the planets revolve at different rates because they are at different distances from the sun. As the whole investigation is of a statistical nature, its results do not apply to individual stars, but only to the average motions of small groups of stars which happen at the present moment to be close together in space. The small group of stars surrounding the sun is found to be describing an orbit with a period of about 250 million years, around a centre of which the direction can be determined with fair accuracy, although its distance offers some difficulties.

THE STRUCTURE OF THE GALACTIC SYSTEM. Whatever this distance may be, there is an obvious inconsistency with Sir William Herschel's conclusion (p. 242) that the sun is at or near the centre of the galactic system, a conclusion which had been accepted well into the present century, and even seemed to be confirmed in 1905 by Kapteyn's statistical studies on the distribution of the stars in space. In brief, the distribution of the stars then seemed to suggest that we are very near to the centre of the galactic system, while the motions of the stars now suggest that we are very far away from it.

This puzzle was resolved about 1920 by the discovery that interstellar space is not perfectly transparent, but contains obscuring matter which impedes the free passage of light, and so limits our range of vision; we live in a sort of cosmic fog. When we walk in a forest on a foggy day, we only see the trees that lie within a certain distance from us, so that we are at the centre of all the trees we see. We must not, however, conclude that we are at the centre of the forest; we are only at the centre of our sphere of visibility. Herschel and Kapteyn, finding that we were at the centre of all the stars we can see, fell into the error of thinking that we were at the centre of the whole system of stars.

Estimates vary as to the dimming power of this cosmic fog, but all are agreed that the fog is thick enough to prevent our seeing the more distant stars of the galactic system; more than half the stars of this system are hidden from our gaze. Yet other objects known as 'globular clusters'—close groups of millions of stars—are bright enough to penetrate through far greater depths of the fog. As these clusters abound in Cepheid variables (p. 349), their distances are easily measured. In 1918, H. Shapley, now the Director of Harvard Observatory, found that roughly they occupy the interior of a disk-shaped space, which lies in or near to the plane of the galaxy. This circle is now known to have a diameter of about 100,000 light-years, and its centre is not at or near the sun, but lies at a distance of about 30,000 light-years from it in roughly the same direction as the centre round which the sun is revolving in its orbit. All our present knowledge is consistent with the two centres being identical, so that the sun (or better, the solar group of stars) must be supposed to revolve round this centre at a distance of about 30,000 light-years, taking 250 million years for a complete revolution, and therefore moving at about 270 km. per second.

The recognition of this cosmic fog has cleared up another difficulty which beset astronomers in the early years of the century. Astronomical objects could be divided into distinct classes by their general appearance, and it was then found that these classes could be divided into two groups, of which one seemed to 'shun' and the other to 'favour' the galactic plane. The latter classes of objects were seen only in parts of the sky which lay near to the galactic plane, the former class only in regions which lay far away from it. We know now that there is no shunning or seeking in any physical sense. But as the cosmical fog is particularly dense and particularly extensive in this plane, some faint objects cannot be seen if they lie in the plane, and so seem to shun it—not because there are none there, but because those which are there cannot be seen through the thick fog. Other classes of objects, such as near bright stars, are hardly troubled by the fog, and seem to seek the galactic plane, because, like most types of object, they are more plentiful here than elsewhere.

THE EXTRA-GALACTIC NEBULAE. Conspicuous among the objects which seem to shun the galactic plane are the 'extra-galactic nebulae', the nebulous-looking objects which Kant and Herschel thought might be systems of stars like our own galactic system (p. 243). They are now known to lie entirely outside the galaxy, and none of them can be seen through the fog near to the galactic plane, because their light is too dim to penetrate it. Those which are visible in other parts of the sky look so faint that they need the largest telescopes of all for their examination.

In 1924, E. Hubble, working with the great 100-inch telescope of Mount Wilson Observatory, found that the outer regions of the most conspicuous of these nebulae, the 'Great Nebula' in Andromeda, could be resolved into innumerable faint stars; it was a discovery analogous to that which Galileo had made when he first turned his telescope on to the Milky Way. Quite recently (1944) Baade has found that the same is true for the inner regions of this nebula. The same procedure has been applied to a great number of other nebulae, and all of them are found to be systems of stars more or less similar to our own.

Most of the nebulae can be arranged in a continuous linear sequence, along which their size and shape change continuously. All nebulae which are of the same shape are assumed to be equal in size, and also of the same intrinsic brightness, so that here again we have types of standard articles—this time the nebulae themselves. Apparent differences in size and brightness must be caused by differences of distance, so that we can measure the relative distances of the nebulae. Hubble had recognised some of the stars in nebulae as Cepheid variables (p. 349) and hence had been able to measure the absolute distances of many of the nebulae, and so to obtain a yardstick for nebular distances.

He and his colleagues have studied the general distribution of the nebulae in space, and find that they are fairly uniformly spaced at average distances apart of about two million light-years. The nearest are at distances of about 700,000 light-years, while the farthest which are visible are perhaps about 1000 million light-years distant. There is no thinning out at great distances, such as Herschel found in our own system of stars, but there is an average uniformity as far as the telescope can reach.

Here and there, however, the nebulae cluster into clearly defined groups. Such clusters can only be kept in being by the gravitational attractions of their members. As the speeds of motion of the members of a cluster can be determined spectroscopically, it is easy to evaluate the gravitational attraction, and hence to weigh the nebulae. Such estimates of nebular masses have been made by Sinclair Smith and others. The average nebular mass usually comes out at some 100,000 million to 200,000 million suns, again showing that the nebulae are star systems like our own.

THE EXPANDING UNIVERSE. The distribution of the nebulae in space hardly raises any fundamentally new problems, except that we should like to know how far it extends beyond the range of our telescopes, but it is different with the motions of the nebulae. Here the theory of relativity is applicable to the situation.

We have seen how Einstein's theory gave a satisfactory account of the motion of the planets round the sun. The discussion of the universe as a whole raises more difficult problems. We have to suppose that all space contains matter, the curvature of the space in any region being determined by the amount of matter present. Each bit of matter adds a bit of curvature to space, and if the matter is everywhere right in amount, its total effect may be just to close the space up into a finite closed volume which will then be in equilibrium. For any assigned density of matter there will only be one size of space consistent with equilibrium.

Einstein had assumed at first that the size of space must be determined in this way. The average density of matter in space could be estimated from the known masses and average distances of the nebulae, so that it was possible to calculate what this size would be. The actual calculations made at the time do not much matter, for they have been superseded, but both they and more recent calculations suggested that the number of nebulae in space must be comparable with the number of stars in a single nebula, which is of the order of 150,000 million. Only about 10 million of these are visible in the largest telescopes.

This relativity picture of a space which was held in equilibrium by the pressure of the matter it contained—rather like a rubber balloon held in equilibrium by the pressure of the gas inside it—seemed satisfactory enough until it was shown by the Russian Friedmann in 1922, and the Belgian Lemaître in 1929, that such an arrangement could not be permanent. It made space into a sort of coiled spring with the curvature impressed by the matter it contained, rather as curvature may be impressed on a spring by laying a weight upon it. If matter moved from one place to another, the curvature would change in both regions, and the universe would no longer be in equilibrium. The new forces thus brought into play might either tend to restore the original equilibrium or to increase the disequilibrium. Friedmann and Lemaître proved that they would do the latter, so that Einstein's proposed arrangement would be one of unstable equilibrium; his space, if left to itself, would start either to expand or to contract.

While all this was still under discussion, Hubble and Humason at Mount Wilson obtained results which, if interpreted in the most obvious way, seemed to show that space was actually expanding and this at no mean rate. The spectra of the nebulae showed displacements which, again interpreted in the most obvious way, suggested that the nebulae were receding from, or advancing towards, the sun. Some of the motion would of course result from the sun's motion of 270 km. per second round the centre of the galaxy. Hubble and Humason found that after this had been allowed for, the remaining displacements indicated that all the nebulae were receding from us at speeds which were proportional to their distances from us. The greatest observed displacement corresponded to a speed of 26,000 miles per second, or about one-seventh of the speed of light.

If the most obvious interpretation was put on all this, space could no longer be compared to the surface of a rubber balloon in equilibrium, but to the surface of a balloon which was being blown out. The nebulae might be compared to small studs imbedded in the rubber, the spectral displacements suggesting a uniform expansion of the distances between our stud and all the others. Superposed on to this systematic motion of expansion were other, and generally smaller, individual motions of the nebulae which conformed to no obvious plan.

If such a motion of expansion is traced back in time, it is found that the whole universe would be confined within a very small volume of space at a period of a few thousand million years ago, a period which is just about comparable with the age of the earth as indicated by its radioactive rocks. This led Lemaître to conjecture that the matter of the universe might be the debris resulting from the explosion of a single big super-molecule. But without adopting so realistic an interpretation as this, we may notice that the apparent motions of the nebulae provide us with a unit of time which is of the order of the age of the earth and probably also of the stars.

This brings us very near to the frontier of present-day knowledge. Reconnaissances have been made into the still unconquered territory beyond, but so far without very tangible or satisfying results. So far the conception of the expanding universe has opened up more questions than it has solved.

This is obviously important; nebular astronomy, the physics of the infinitely great, is seen to tell the same story as radioactivity, the physics of the infinitesimally small, and physics is confirmed to be a consistent whole.

Sir Arthur Eddington devoted the last years of his life to an attempt to establish a yet wider synthesis which should connect fundamental facts of different sciences, and show that they were all necessary consequences of certain basic assumptions, which seem to have been the fundamental principles of the quantum-theory, heavily camouflaged.

On this as basis, he claimed to have shown that the speeds of nebular recession must of necessity be just about those actually observed; that if we picture the universe as a structure in space and time, then the number of dimensions of space must necessarily be three, and of time one; that if we picture the universe as consisting of particles, some will be negatively and some positively charged; that the ratio of the masses of the two kinds of particles will be the ratio of the roots of the quadratic equation $10x^2-136x+1=0$, or 1847·6, a number which, whether rightly or wrongly obtained, is certainly very near to the observed ratio of the masses of the proton and electron; that the total number of particles in the universe is necessarily

$$\frac{3}{2}\times 2^{256}\times 136,$$

and that this is a simple assigned multiple of the square of the ratio of the electrical and of the gravitational attractions between an electron and a proton.

Few, if any, of Eddington's colleagues accepted his views in their entirety; indeed few if any claimed to understand them. But his general train of thought does not seem unreasonable in itself, and it seems likely that some such vast synthesis may in time explain the nature of the world we live in, even though the time may not be yet.

INDEX

Printed in the United States
By Bookmasters